T0091713

SEEKING ASYLUM

SEEKING ASYLUM

*Human Smuggling and Bureaucracy
at the Border*

Alison Mountz

University of Minnesota Press
Minneapolis
London

Portions of chapter 2 were originally published in "Human Smuggling and the Canadian State," *Canadian Foreign Policy* 13, no. 1 (2006): 59–80.

Published by the University of Minnesota Press
111 Third Avenue South, Suite 290
Minneapolis, MN 55401-2520
http://www.upress.umn.edu

Library of Congress Cataloging-in-Publication Data
Mountz, Alison.
Seeking asylum : human smuggling and bureaucracy at the border /
Alison Mountz.
      p. cm.
   Includes bibliographical references and index.
   ISBN 978-0-8166-6537-2 (hc : alk. paper)
   ISBN 978-0-8166-6538-9 (pb : alk. paper)
1. Human smuggling.   2. Emigration and immigration.   3. Human
smuggling—Prevention.   4. Illegal aliens.   I. Title.
   JV6201.M68 2010
   364.1'37—dc22

Printed in the United States of America on acid-free paper
The University of Minnesota is an equal-opportunity educator and employer.

18 17 16 15 14 13 12 11 10   10 9 8 7 6 5 4 3 2 1

I see a whole generation
freefalling toward a borderless future . . .
I see them all
wandering around
a continent without a name . . .
all passing through Califas
enroute to other selves
& other geographies
(I speak in tongues)
standing on the map of my political desires
I toast to a borderless future

—Guillermo Gómez-Peña, *The New World Border*

Contents

Acknowledgments

THIS BOOK WOULD not have been possible without the support of many people in many places: British Columbia, Ottawa, Hong Kong, New York, New Jersey, Mexico, El Salvador, and Australia. It is risky to open oneself to reflective discussions about life and work, and I am humbled by the generosity of people in the field who provided insight and became friends as much as participants in my research, and who taught me about migration and so much more. They care deeply about these issues for reasons both personal and professional. They are not named in the text due to the precarious nature of their legal status and the sensitive nature of their work.

In British Columbia, I am especially grateful to Dan Grant, Chris Taylor, Alex Charlton, Joshua Sohn, Robin Pike, and Meyer Burstein. A broad group of dedicated activists and people working at various institutions and organizations also gave time to this project. I am indebted to the many people from Mexico, El Salvador, and China who spoke with me in Canada and the United States about their journeys. Although they may not feature prominently in this text, they are a central part of the motivation behind its existence.

At the University of Minnesota Press, I thank editor Jason Weidemann, who is patient and encouraging, editorial assistant Danielle Kasprzak, who was very helpful, and many others who worked hard to complete this book.

I owe a great intellectual debt to mentors and advisors: Richard Wright, David Ley, Gerry Pratt, Dan Hiebert, Vicky Lawson, Adrian Bailey, and Ines Miyares.

Colleagues farther away provided intellectual camaraderie and dialogue at conferences and via e-mail, especially Steve Herbert, Mat Coleman, Juanita Sundberg, Susan Coutin, Catherine Dauvergne, Lynn Staeheli, and Valerie Preston. My Syracuse writing group read chapters: Margaret Himley, Robin Riley, and Jan Cohen-Cruz. Thanks to all.

Research was supported generously by the John D. and Catherine T. MacArthur Foundation, the Canadian Embassy, the Killam Foundation, the University of British Columbia Faculty of Graduate Studies, Metropolis of British Columbia, the National Science Foundation, Dartmouth College, the Canada Program of the Weatherhead Center for International Affairs at Harvard University, and the Helen Riaboff Whiteley Center at Friday Harbour Laboratories.

At Syracuse University, I appreciate support from the Department of Geography, the Maxwell School, and Trish Lowney at the Office of Sponsored Programs. Special thanks go to cartographer Joe Stoll for illustrative maps, beautiful photographs, and to Jake Bendix for strong coffee. I am grateful to administrative staffs at the departments of geography at the University of British Columbia and Syracuse University, whose daily support made it possible to keep going, especially Chris, Janet, and Jackie in Syracuse and Junnie, Jeannie, Elaine, and Lorna in Vancouver.

Departmental colleagues in Syracuse engaged this work in different ways, including Mark Monmonier, Bob Wilson, Jenna Loyd, Tod Rutherford, Don Mitchell, Anne Mosher, John Western, and Jamie Winders. Several students in Syracuse were talented research assistants on this project: Eunyoung Choi, Kate Coddington Senner, Ishan Ashutosh, Nancy Hiemstra, Jackie Micieli, Caro Harper, and Joaquin Villanueva.

Many friends have supported, distracted, entertained, and laughed with me over the years: Helen Watkins, Win Curran, Beverley Mullings, Shelly Rayback, Margaret Walton-Roberts, Jackie Orr, Jen England, Millie Rossman, Maija Heimo, Rowan Flad, In Paik, Kevan Barton, Geoff Rempel, Graham Webber, Natalie Oswin, Wendy Mendes, Magdalena Rucker, Caroline Desbiens, Bonnie Kaserman, Minelle Mahtani, Lori Brown and Martin Hogue, Samantha and Michael Herrick, and Ruth and Sue La Porta.

Finally, I thank my family—Henrietta, Robert, Sarah Mountz, the other Mountzes, the Tulimieris, and most of all Jennifer Hyndman, who taught me about love and belonging.

Abbreviations

ALO	Airline Liaison Officer
BC	British Columbia
CBSA	Canada Border Services Agency
CIC	Citizenship and Immigration Canada
DFAIT	Department of Foreign Affairs and International Trade
DHS	Department of Homeland Security
EU	European Union
ICO	Immigration Control Officer
INS	Immigration and Naturalization Services of the United States
IOM	International Organization for Migration
IRB	Immigration and Refugee Board of Canada
MRT	Marine Response Team of CIC
NGO	Nongovernmental organization
NHQ	National Headquarters, Citizenship and Immigration Canada
POE	Port of entry
PRC	People's Republic of China
RCMP	Royal Canadian Mounted Police
RHQ	Regional Headquarters, Citizenship and Immigration Canada
STCA	Safe Third Country Agreement
UNHCR	United Nations High Commissioner for Refugees

Introduction   **Struggles to Land in States of Migration**

Openings and Closures: "The Long Tunnel Thesis"

On July 20, 1999, off the coast of British Columbia, Canadian authorities intercepted what would be the first of four boats to arrive during a period of six weeks with a total of 599 tired and hungry women, men, and children on board. The *Yuan Yee* carried 123 people from the coastal province of Fujian, China. They were estimated to have been at sea for approximately thirty-nine days, smuggled on retrofitted fishing trawlers from Fujian and intercepted en route to North America. The Department of National Defense (DND) first spotted the boat as it entered Canadian territorial waters, which begin and end twelve miles offshore. The DND contacted the Department of Citizenship and Immigration Canada (CIC), the federal agency tasked with managing immigration and refugee claims and with border enforcement. Ultimately, some twelve federal departments became involved in the interceptions, but CIC served as coordinating agency in the interceptions and subsequent processing and detention. A harrowing encounter ensued after this and each subsequent ship was intercepted. Federal ships chased smugglers, who tried to conceal their cargo as well as the details of the journey. They jettisoned GPS and other technology that would betray the details of what had by then become a highly lucrative industry of transporting migrants across the Pacific to North America. British Columbia proved a desirable entry point, where smugglers believed the ships could land undetected along the isolated, if rugged, coast. Though it was not yet known, the migrants were most likely headed south to the United States.

After being transported for processing to the Esquimalt naval base outside of Victoria on Vancouver Island, most made refugee claims, asking the Canadian government for protection from persecution they might suffer if returned home.[1] Once a person makes a refugee claim in Canada, he or she is generally granted access to a refugee lawyer and begins the challenging task of proving—like refugee claimants everywhere—that his or her fear of persecution, if returned home, is well founded. This did not happen right away in the case of those intercepted from China, so Canadian refugee lawyers actually had to request access to the claimants. A heated debate ensued. The government wanted to buy time to learn as much as possible about the migrants and the conditions of their journey, conduct a criminal investigation, and figure out how to proceed with processing. In response to the lawyers' request, the government reworked the local geography by temporarily designating the base a "port of entry," thereby declaring the migrants in the stage of being processed. This decision shaped migrants' access to legal rights. Authorities denied migrants access to lawyers until processing in the port of entry had been completed. During this time, immigration officers conducted preliminary interviews with migrants to determine their intentions in landing in Canada. These interviews and the information collected later became the source of controversy and legal struggle.

As I conducted research with the Department of Citizenship and Immigration Canada in the aftermath of the boat arrivals, officials at times proudly and at times jokingly explained to me the details of this decision that they called "the long tunnel thesis." Migrants would be treated as though they were walking through the long tunnels of an international airport, not yet landed on Canadian soil. This decision transformed sovereign space for a window of time into what Suvendrini Perera (2002c) would call "not Canada." Although the migrants were located on Canadian sovereign territory for the duration of their processing, they were not yet in Canada for legal purposes. Instead, they found themselves in an interstitial processing zone, somewhere between Canada and not-Canada, paradoxically neither in nor out.

This book explores what states do in their interventions into human migration. What are their daily practices regarding populations on the move? How do civil servants working under stressful conditions of crisis respond to human smuggling? Where does this work happen? The story above illustrates but one of many struggles between states and migrants over migration and refugee policies. The brief yet important intervention in western Canada when bureaucrats designed the "long tunnel thesis" signals the importance of locating decisions about migration with attention to the microlevel, grounded daily practices of government employees working simultaneously in urban office towers and along remote borders of Canada. The elongation of the long tunnel raises questions about the power of states to alter the relationship between geography and the law. Migrants struggle to land on sovereign territory to work or to access the claimant process, and states alter time and space in response, manipulating geography in times of crisis. The "long tunnel thesis" proved only one small part of a broader constellation of enforcement practices taking place globally. States of migration can restrict access to those attempting to enter illicitly, either to work or to make a refugee claim. They use a range of tools to capture people who are in various states of migration. The ground beneath shifts as one moves through time and space. Altered geographies such as the long tunnel reflect the changing spatiality of sovereign power and the corresponding dimensions of exclusion.

This book argues that immigrant- and refugee-receiving states capitalize on the crises generated by high-profile human smuggling events to implement a series of restrictive measures designed to control migration. I make this argument by documenting the daily practices of the state in its work on migration. The "long tunnel" provides entry into discussion of the main themes of this book, introduced in this chapter and elaborated in the chapters that follow.

In times of crisis, an air of exceptionalism prevails (Agamben 1998; Klein 2006). Human smuggling is in many ways routine, similar to other kinds of migration. Human smuggling is also a historical phenomenon, one that has existed for centuries; for as long as nation-states have asserted control of mobility along their borders, people have employed assistance to cross them. Yet highly visible interceptions of smuggling operations tend to play out as contemporary crises. Due to the sensationalism of stories about smuggling and human interest in the securitization of national borders, the media partakes in and promulgates these events as crises. Escalated media coverage heightens public fears about sovereign control of migration. These security crises enable officials to embark on seemingly exceptional enforcement agendas, from detention in remote locations where access to legal representation is limited, to the design and implementation of more restrictive, "ad hoc" policies. Migrants and potential refugees struggle in their efforts not only to disembark, but to then achieve access to sovereign soil and the human and legal rights attached to sovereign territory.

Resistance to developing long-term migration strategies fosters crises when human smugglers transport migrants illicitly across borders. In the ensuing government responses, policies spring up reactively. In western Canada, for example, regional bureaucrats referred to their response to human smuggling during the boat arrivals as "policy on the fly." Australians similarly referenced their government's response to human smuggling by sea in recent years as "policy on the run." It is essential to understand how human smuggling is produced through these narratives of exceptionalism in order to test exceptionalism as an intricacy of exclusion.

Commentaries on globalization throughout the 1990s referred to an emasculated state, penetrated by global flows of migrants and capital, transnational organized crime, and foreign direct investment (Chang and Ling 2000). Yet states asserted themselves strategically in these industries. They negotiated free trade arrangements, recruited the wealthiest or most skilled "flexible citizens" (Ong 1999) and "millionaire migrants" (Ley 2010), and even developed symbiotic relationships to "global" crimes with heavy investments in border-policing industries (Nevins 2002). Through more transnational border enforcement practices, states grew more dispersed, though no less powerful (Mountz 2006; Coleman 2007).

States develop narratives to explain and perform their day-to-day work. In my research, several narratives emerged. One narrative involves a powerful state, able to flex its muscles to exclude through enforcement. A second narrative is that of the vulnerable state, often repeated by civil servants who themselves feel underresourced and ill-informed about human smuggling, and who confront daily the frailties and contradictions of public policies. These contradictory accounts belie times of "crisis" when they become most apparent, most intense, most at odds, and therefore most productive. States perform narratives, and, amid the dissonance, excel in particular at performing crisis. In crisis, nation-states pit protection of their own citizens against a broader commitment to protect human rights. Contradictory projects are carried out daily in the name of vague security agendas.

Immigrant-receiving states of the global North do not operate in a vacuum of exceptionalism. On the contrary, they look to peers for "best practices." Throughout the 1990s, as high-profile human smuggling episodes increased in North America, Australia, and Europe, the securitization of migration was well underway (Huysmans 2006), alongside a corresponding criminalization of smuggling. In cyclical fashion, as states increase enforcement abroad, smuggling enterprises benefit and grow (Nadig 2002). These enterprises generally carry what policymakers refer to as "mixed flows" of economic migrants, moving primarily to find work, and would-be refugees, who might fit the rough contours of a definition laid out by the 1951 Convention Relating to the Status of Refugees. As a result, many people in search of asylum or refugee status fall victim to increased border enforcement abroad and never reach sovereign territory to make a claim.

The crisis-driven discourse is accompanied by a corresponding geography where states manipulate space to subvert access to human rights, alter legislation, and extend their reach beyond territorial borders. These strategies are part of the securitization of migration. This entails processes wherein people crossing borders find themselves subjected to heightened surveillance and enforcement in the name of national security. Increasingly, states exercise transnational enforcement practices such as detention and interdiction, keeping asylum seekers far away from sovereign territory. As a result, crises often transpire along the geographical margins of sovereign territory: on islands, in airports, at sea, and in offshore detention centers where authorities and migrants encounter each other. These sites are hypervisible during crises and yet simultaneously obscured from view of the general public and human rights monitors. This paradoxical location makes them ideal sites for enforcement activities where states manipulate geography.

States have been shrinking spaces of asylum through creative enforcement in peripheral zones. Authorities work well beyond land crossings and coastal borders of sovereign territory, reaching outward to stop migrants en route. Through this process, the border moves, only to be reconstituted around the bodies of refugee claimants. In the long tunnel, for instance, a dynamic threshold to Canada was created, where the border was actually moved to the sites where claimants were intercepted and then processed. As a result of more dispersed enforcement practices, the statistics of successful asylum applications diminished quietly in the 1990s and then dramatically following 11 September 2001 (Newland 2005; UNHCR 2007). Negative discourse about those smuggled accompanies interceptions and also functions to shrink opportunities for asylum.

Canada led the way among states enacting creative enforcement practices, beginning in the 1980s with its "multiple borders strategy," wherein borders were policed offshore before migrants ever approached Canadian sovereign territory. Canada thus serves as both leader and exemplar in these practices, illustrated in this book with an ethnography of its Department of Citizenship and Immigration Canada in chapters 1 to 4, followed by international examples in chapters 5 to 7.

Negative discourses about groups of people produce identities that accompany exclusionary geographies. As I researched the events that had unfolded at the Esquimalt naval base in western Canada, the Halliburton corporation was busy preparing Camp X-Ray at Guantánamo Bay on behalf of the U.S. government to detain "enemy combatants" taken as prisoners during U.S. military action abroad. A heated debate ensued in the years to come about the legal status of detainees and their access to legal representation. They too were denied access to lawyers and to U.S. courts because of their location.

In Australia, conservative Prime Minister John Howard ran a successful reelection campaign in 2001 on an anti-immigrant platform, capitalizing on fears generated by politicized and publicized interceptions of boats carrying smuggled migrants north of Australia. Howard successfully implemented retroactive legislation that declared hundreds of islands off the coast of Australia no longer part of Australian territory for the purposes of migration. This strategy, known as "the power of excision," effectively disrupted people's attempts to reach sovereign territory by boat to make an asylum claim. A few years later, Italy pursued a related strategy, altering its response to migrants traveling by boat from Africa to the shores of its small island of Lampedusa. No longer willing to transfer refugee claimants to a reception center for processing on Sicily, Italian authorities shifted policy to aggressive detention and return of migrants en masse (Rutvica 2006). These struggles continue along the borders of North America, Australia, and the EU.

Geography and the law are intertwined in many ways, and legal identities of migrants take shape through the production of particular geographies. Legal identities of those "at sea" or "detained abroad" correspond with an assumption about who "they" are. The media advance a discourse of the "bogus refugee," a totalizing narrative that assumes a priori that claims for asylum will prove false and turns public sentiment against public policy designed with the initial goal of protection. This production of legal identity corresponds with access: those scripted as "bogus refugees" before their claims are heard may face restricted access to the claimant system. Among the victims of crises are migrants in search of asylum as well as those bureaucrats working "the front lines" to implement policies.

The juxtaposition of refugee resettlement policies that protect and border enforcement practices that exclude highlights the relationship between geography and the law. The states involved in offshore management of approaching migrants are all signatories to the 1951 Convention Relating to the Status of Refugees and its 1967 Protocol, and among those with more established immigration and refugee resettlement policies. They are also those with the most sophisticated border enforcement, detention, and deportation regimes. Actions taken abroad to protect and resettle refugees contradict offshore activities to stop asylum seekers from reaching the sovereign territory of these signatory countries. In each of the aforementioned cases, authorities acted in remote locations and were subsequently accused of undermining commitments to the Convention and its Protocol. They did so by extending their operations beyond the borders of sovereign territory, where the legality of actions becomes ambiguous. No longer confined to the compact territories demarcated by international borders, states reconfigured geographies of exclusion.

Nation-states thus prove themselves to be transnational actors in relation to international migration. The dimensions of this geography emerge not necessarily in the texts of government policy, but in the narratives of migrants and civil servants who embody its policies.

Enter the Ethnographer

I approached the nation-state as a subject of analysis as the result of work with immigrants and asylum seekers. For those on the move, the state is a powerful, dynamic entity, constantly in motion. It holds the power to inhibit and otherwise structure human mobility through material barriers as well as imagined geographies where people reproduce state boundaries in everyday life. This text, therefore, aims to construct the state through narratives of those who produce its borders daily. Some voices belong to civil servants who struggle to enact migration policies in their daily work, while others belong to people who struggle with state policies in every facet of their daily lives, because they have not been granted the safety of legal belonging.

After conducting ethnographic research with undocumented Mexican and Salvadoran migrants and asylum seekers in the United States in the mid-1990s, I had been circling around the state as I watched the effects of policies reverberate through the daily lives of those without citizenship. I decided to move my research "inside" the state, using ethnography to understand how federal bureaucracies responded to a variety of transnational migrations. This desire brought me to a highly controversial episode in recent Canadian history: the interception of the four boats carrying migrants from Fujian, China, to British Columbia, Canada. In August 1998, only one year before many of the migrants came by boat, I had arrived in

Vancouver from New York. In the wake of these arrivals and the ensuing detentions and legal struggles, I conducted institutional research on, and spent months in, Citizenship and Immigration Canada, the federal bureaucracy tasked at the time with immigration and refugee policy and border enforcement.[2]

In order to explore the role of the nation-state in ordering human migration, I examined government responses to human smuggling by employing qualitative research methods within the machinery of the immigration bureaucracy. Methods included ethnography to study the daily work of bureaucrats responsible for implementing policy; archival research on documents pertaining to interceptions; and semistructured interviews with nongovernmental people, including media workers, refugee lawyers, advocates, activists, refugee resettlement workers, and immigrants.[3] Research took place primarily at CIC and included about seventy interviews not only with CIC employees, but also with people in many other locations.[4] Ethnographic material provides an opportunity to test, qualify, and advance theories and to connect daily, localized activities to global forces. Seemingly mundane decisions in the daily work of bureaucrats entangle global forces and localized institutional arrangements.

The summer of 1999, when the boats were intercepted, turned out to be a season that tested the standard operations of procedure and policy through which government bodies interact and marshal resources. The response drew together many different institutions, working through distinct mandates and cultures. It revealed the disparate ways in which civil servants and human smugglers work along and across international borders. I found the state to be as transnational as those migrant communities where I had worked previously.

The research contributes the voices of civil servants to conceptual models of immigration (e.g., Heyman 1995; Hyndman 2000; Nevins 2002) and juxtaposes them with the voices of employees of other institutions. Further, my findings reveal that there are competing narratives of state policies and practices. The "long tunnel thesis" was not written in any public policy, recorded in the daily newspaper, or released to the public in any request that called upon the Access to Information Act in Canada. Rather, the thesis was created in daily practice, not as policy, and committed to the institutional memories of a select group of mid-level regional managers and employees. Their knowledge was imparted through interviews and observation in the daily recollections of what was, by all accounts, a chaotic time. This group worked together in a specific time and place; many then went on to other jobs, their institutional memories dispersed with them. As I traced these stories, the ethnography of the state moved well beyond the boundaries of sovereign territory and bureaucratic parameters. Like people migrating between states, bureaucrats act on geographical imaginations that simultaneously reproduce and transcend international borders.

Ethnography of the State

In addition to functioning as a set of institutions on the ground, the state has a lively conceptual existence in the social sciences, generally referred to as theories of the state. Social scientists from a variety of fields offer approaches to understanding the daily work of powerful institutions that shape our lives. Much social science has conceptualized states as containers of people, economic systems, cultural practices, and military power, a phenomenon that John Agnew calls "the territorial trap" (1994, 1999). Yet the contemporary nature of enforcement practices defies geographical confinement even to what political geographers have traditionally mapped as sovereign territory. The powerful machinations of states appear not only in the borders drawn on maps and the pages of public policies, but in the fractured fault lines of daily practice. It is important, therefore, to examine daily life as one register of state power. While countless epistemologies and methodologies exist to conceptualize the state, ethnography proves most effective in tracing its embodiment in the daily lives of bureaucrats and migrants, operating across a dispersed spectrum of geographical locations that Akhil Gupta (1995, 392) calls the "translocalities" of the state. Here I review just a few key strands at work in these literatures that assist in tracing the contours of ethnography of the state.

Too many theories of the state fail to embody state practices enacted in daily life. Such failures to locate "the state" reify its mythical quality. The disembodied state remains ubiquitous, existing everywhere and nowhere. As Phillip Abrams suggests, "[T]he state, conceived of as a substantial entity separate from society has proved a remarkably elusive object of analysis" (1988, 61). Likewise, Timothy Mitchell argues that we view states with "aridity and mystification rather than understanding and warranted knowledge" (1991, 61). In much scholarship in the social sciences, the state has been elusive, "out there" (Mitchell 1991, 94), mythical by nature (Hansen and Stepputat 2001, 20–21; 2005), a "timeless national essence" (Steinmetz 1999, 12). "We are variously urged to respect the state, or smash the state or study the state: but for want of clarity about the nature of the state such projects remain beset with difficulties" (Abrams 1988,

59). Mitchell notes that scholarship falls prey to what he calls "the structural effect": the artificial division between state and civil society (1991, 94). The "structural effect" reinforces the notion that the state is beyond our control. Located nowhere in particular, the disembodied state is imagined as unitary, omniscient, magical (Taussig 1997). Disembodied depictions depoliticize and disempower as communications strategists craft a unified, coherent narrative for public consumption.

Scholars made premature declarations regarding the "death of the state" in an era of intensified global flows (e.g., Ohmae 1995; Appadurai 1996). Indeed, following the "cultural turn" and influenced by Michel Foucault (1991; 1995), social scientists have grown less interested in the state and more invested in conceptualizing power as dispersed within and beyond institutions (Steinmetz 1999). Exercises in state power shifted, according to Foucault, from sovereign to disciplinary power. The architecture of institutions migrated into and reverberated through the social body, where identities were produced and policed. Many contemporary scholars of citizenship concern themselves with the latter, looking particularly to the power of surveillance in its regulation of persons through the collection of demographic data (Amoore 2006). A robust literature on Foucault's (1991) idea of governmentality, or the "conduct of conduct," rather than governance has emerged (e.g., Dean 1999). Whereas states once asserted power through repression, this literature takes into account the power of the state to operate in productive ways, producing subjectivities through the collection and ordering of information, practices, and identities. Although the neoliberal state has been scaled back, hollowed out, and devolved through public and private contracts, it is hardly defunct. Studies of the more sinister effects of neoliberal policies took hold, even as the neoliberal state retreated and retrenched. Interest grew in the internationalization of states and in devolution, or the role of the shadow state (e.g., Wolch 1989). Some scholars argued that the liberal welfare state was being hollowed out as a result of global flows; others looked at its more scattered reconfiguration, as diasporic populations and global networks facilitate the reterritorialization of communities and the structures that govern them (e.g., Basch et al. 1994; Appadurai 2006).

Working across fields of anthropology, sociology, geography, and political science, authors have converged on the epistemological and methodological significance of the everyday dimensions of states. Laura Nader (1972) called anthropological study of institutions "studying up." Studying up offers starting points to resist projects of the state, however modest. Dorothy Smith (1987, 1990) mapped out a related field termed "institutional ethnography" to trace institutional power's movement outward through what she calls the "relations of the ruling." Among the latest incarnations of studying up to understand institutions is a growing field called "anthropology of the state" (Sharma and Gupta 2006). Joining in these traditions, this book asks what it means to tell day-to-day stories of the state in relation to the experiences of transnational migrants. Like the quest to "ground" globalization (e.g., Burawoy 2000; Smith 2001), this ethnography of the state joins this recent journey to know the state in new ways, to locate where and how state powers and policies function on a daily basis.[5]

Ethnography of the state offers potential to counter disembodied theories by recording dispersed sovereign powers in daily practice. No longer the tool of the lone anthropologist studying "ethnic" rituals in remote villages of people speaking obscure languages, ethnography has taken on new meanings and critical practices that include examining our own institutional practices in recent years (Clifford and Marcus 1986; Gordon and Behar 1995). Contemporary ethnographers endeavor to record daily life. Ethnography of the state uncovers the operation of power at multiple scales and centers; it appreciates the locus of power as perpetually in motion and, in this book, bound geographically with transnational flows of migrants and information.

The study of states as daily, embodied entities connects narratives that structure the routine work of bureaucrats with human migration. The ethnography outlined in chapters 1 through 4, for example, documents the conflicting narratives at work in times of crisis, demonstrating how bureaucrats see human smuggling and work to implement policy in the midst of crisis. International examples in subsequent chapters show that Canada did not act in isolation. Crises brought on by the very public images of human smuggling played out on European Mediterranean Islands, Australian Pacific Islands, and North American coastal and land borders throughout the 1990s and continue to do so today, prompting and enabling these states to respond with more aggressive exclusionary practices.

Mapped ethnographically, the state is an idea that is imagined, shared, and performed by a set of institutional actors with powerful material consequences, including new spatial dimensions of exclusion. The state functions as a set of networks (Painter 1995; Nyers 2006) composed of individuals whose interactions cross international borders as well as the boundaries of institutions. According to Painter (1995, 34), "States are constituted of spatialized social practices which are to a greater or lesser extent institutionalized (in a 'state apparatus')."

The state operates not as a locatable object per se, but as a located series of networks through which governance takes place. "The state" does not exist outside of the people who comprise it, their everyday work, and their social embeddedness in local relationships. As Mitchell (1991, 93) argued:

> [J]ust as we must abandon the image of the state as a free-standing agent issuing orders, we need to question the traditional figure of resistance as a subject who stands outside the state and refuses its demands. Political subjects and their modes of resistance are formed as much within the organizational terrain we call the state, rather than in some wholly exterior social space.

Mitchell's analysis conjures a spatial paradigm in which social scientists reify the artificial divide by theorizing the state and civil society as separate entities (Mitchell 1991; Gupta 1995; Kalyvas 2002). Rather than the repressive, autonomous body that affects social relations, the state is itself a set of daily practices (e.g., Herbert 1997; Mountz 2003).

For Joe Painter (2006, 752), states must be understood in "prosaic" ways, meaning that "everyday life is permeated by stateness in various guises." Conceptualization of states as daily practices opens political opportunities for resistance to state projects, in part by showing the inroads of civic groups, nongovernmental organizations, and legal advocates. These institutions sometimes support and other times challenge the work of the state through devolution. The hollowed-out neoliberal state has less capacity, for example, to do the day-to-day labor of refugee resettlement and contracts out this labor to nongovernmental refugee resettlement organizations. Still other organizations do the work of legal advocacy, job training, and language instruction. In this way, work in the field of immigration challenges easy divisions between state and non-state actors, policy and practice, and state and civil society. A variety of people work across institutions in collaboration, collusion, conflict, and contradiction.

Daily struggles breathe life into unitary or monolithic conceptions of the state. Through daily work, civil servants confront structural constraints as they attempt simultaneously to subvert them (Herbert 1997). As narratives of border management show, a profound change in an individual decisionmaker can be a profound change within the state. In this way, immigration officers have the potential to be subversive in their day-to-day work, particularly when a critical mass begins to question and challenge policy. Ethnographic observation, through its documentation of these frustrations, subversions, and networks among governmental and nongovernmental players, can contribute to political breaking points within the state, theorized as an institutional arrangement of social practices.

The embodied nation-state is articulated through, and in an important sense limited by, the imaginations of those who enact it. Behind each decision are individuals acting within varied institutional and geographical contexts. Many state theories, however, do not locate the nation-state geographically in a time or a place, but rather assume its pervasive and homogeneous nature (Gupta 1995). Yet state practices are constituted within and through social networks. Embodiment thus serves as a strategy to locate knowledge and power in a time and a place (see Haraway 1988; Gupta 1995; Rose 1995). This strategy deconstructs the monolithic, hierarchical state and locates migrants, refugee claimants, and institutional actors in relation to one another. What began as a localized response to human smuggling in BC eventually led me to Hong Kong and to understand the traveling and haunting nature of state practices that reach beyond geographical bounds.

Ethnography here counters the depoliticizing (Nevins 2002) abstractions of states with a case study of CIC. The practices and beliefs of civil servants can be captured with ethnographic research that demystifies the power of the state. Such work uncovers taken-for-granted assumptions behind public policy. Canadian civil servants, for example, articulated the ways their views on, and roles in, the response to human smuggling worked both symbiotically and in tension with policy. Resistance, therefore, need not come exclusively from "the outside" (see Gupta 1995, 394). The recovery of alternative narratives of the state, sometimes suppressed by the bureaucracy, disrupt some of the more audible narratives about transnational migration that have become normalized as culturally acceptable narratives—e.g., the state as facilitator of capital flows for investment and economic growth— and expose inconsistencies in Canada's self-imaginings. Coherent national narratives often fall apart upon inspection and enable the conjuring of new transnational imaginaries. Within the intricate and intimate connections among institutional subjects lies potential for social change.

With important exceptions (Calavita 1992; Heyman 1995; Nevins 2002; Maril 2004), little has been written about the work of federal immigration officers in the United States, and still less about their work in Canada (but see Foster 1998; Ley 2003; Pratt 2005; Pratt and Thompson 2008). Immigration scholars often discuss "the state" as though government were a unified decisionmaker. But the term does not adequately convey the negotiations over resources and differing mandates that take place within and among governmental agencies. All levels of government may play a role in the governance of migration and in the realm of enforcement, but the federal government holds most responsibility and power in terms of the implementation of policy. Federal bureaucracies play conflicting roles to enforce borders and facilitate immigration. Furthermore, many different kinds of nation-states play a role in processes and politics of global migration. Some are "source" or "transit" countries, others "receiving" countries that intervene.

Immigration policies are not ad hoc, but rather the strategic positioning of groups of people in relation to the global economy through their identification as particular types of transnational subjects. One's identity and location are central to this process of classification, constituted through federal immigration policy. Civil servants who implement immigration and refugee policies come to see transnational migrants in different ways according to various axes of difference and location. Like migrants, bureaucrats experience the world in distinct ways in relation to their distinct social locations in the world, and so relate to different immigrants in different ways. These practices materialize within the bureaucracy, where diverse institutional subjects operationalize policies. They actively define themselves and the nation-state in relation to those whose entrance they facilitate or prohibit. Cultural identities are constructed through quotidian practices, influenced by departmental cultures and the larger body politic (e.g., Heyman 1995; Herbert 1997; Nelson 1999). It is important, therefore, to contemplate the social relations within which the nation-state is enacted by examining interfaces between discourse and materiality (Painter 1995), between institutional contexts of reception, the language of categorization, and actual access to the nation-state (see Sharma 2001).[6]

Much important research exists on the ways that immigrants and immigrant communities experience immigration policy (e.g., Mahler 1995; Chin 1999; Coutin 2000). This, however, is not an ethnography of migrants' lives, although stories are told where appropriate. Ethnography of the state offers a different, but essential, piece of the complex equation of human migration wherein people cross borders for a vast array of reasons, and with varying degrees of success. "Studying up" on the daily practices of the many institutions that regulate migration offers the possibility of understanding the local intricacies and global intimacies of border crossings and the exceptionalism that accompanies them.

Human Smuggling: Exceptionalism and "the Other"

One of the key aims of this ethnography of the state is to challenge the production of exceptionalism as a discursive, geographical, administrative, linguistic, and political project. The labeling—and subsequent exclusion—of certain movements of people as human smuggling demonstrates exceptionalism at work. Human smuggling is but one form of migration, one aspect of continuous, historical journeying where people employ others' assistance to move. The term "human smuggling" marks a category and, like any other category, produces particular identities. In so doing, it reproduces the power of the state through simultaneous inclusion and exclusion (see Honig 2001). The illicit nature of this particular industry draws public attention and enforcement resources in equal measure to one dimension of longer transnational journeys to enable people to work, live, or make claims for asylum. Those who employ smugglers fall under a range of immigration and refugee policies, but are all homogeneously produced as criminal, their access to legal avenues inhibited. These practices illuminate the relationship between discourse and geography, smuggling and remote detention. If migrants have the resources to enter quietly—and therefore less visibly—through airports, they are less likely to be intercepted, their arrival will be less publicized, and their access will be less restricted.

This production of human smuggling ties visuality to the criminalization of migration, enabling the securitization of border crossings through enhanced enforcement. State and smuggler exist symbiotically, each necessitating and hailing the other into being. Transit and destination states of migration reach out transnationally to enforce—and in so doing, reinforce—smuggling industries.

People, meanwhile, do not imagine their lives or identities in the terms of immigration policies and the categories they produce. But the state bureaucrat must imagine his or her work in terms of categories of people, managed by a range of policies, however complicated this becomes.

The language of migration is embedded within political struggles, where the terminology of human smuggling is contested.[7] The media labeled the migrants "queue jumpers," "bogus refugees," and "boat migrants." Here they are referred to as *refugee claimants* because they made claims upon arrival, and as *migrants* because they did not have legal opportunities or the financial means to come to Canada to settle in the long term as immigrants. Whereas human smuggling is defined as "the illicit movement of people across international boundaries" (Koser 2001, 59), "human trafficking" entails coercion, often involving the particularly exploitive niches of the global sex trade and child labor. The UN Protocols on Human Smuggling and Trafficking (United Nations 2000) identify coercion and exploitation on both ends of the journey as the main elements that distinguish trafficking from smuggling. Whereas coercion involves people moving and working against their will, exploitation means making them work in poor conditions for substandard wages. The distinction between smuggling and trafficking is more ambiguous and complex than is often represented, however. Smuggling is often posited as a more entrepreneurial, transactional exchange of services, whereas those subject to trafficking are typically conceived of as "victims" or "survivors" (see Bhabha 2005; Sharma 2005). As Bhabha notes, "the distinction between trafficking and smuggling is difficult to implement in practice" (2005, 4).

This discord between policy and practice rests at the center of negotiations between states and migrants and the many contradictions therein. Migrants struggle to fit themselves into policies, and the mismatch plays out as state reads body and bodies read states.

The Italian philosopher Giorgio Agamben (1998, 2005), perhaps more than any other contemporary scholar, has theorized exceptionalism in marginal zones of the state. His ideas have received much scholarly attention for the sophistication with which he captures the contradictions of sovereign power and paradoxical zones that he renders at once internal and external to sovereign power:

> The sovereign exception is the fundamental localization, which does not limit itself to distinguishing what is inside from what is outside but instead traces a threshold (state of exception) between the two, on the basis of which outside and inside, the normal situation and chaos, enter into those complex topological relations that make the validity of the juridical order possible. (1998, 19)

According to Agamben, paradoxical inclusions and exclusions constitute the state of exception along the margins of sovereign territory, and the work done through these "complex topological relations" in fact fuels governance of belonging and not belonging through citizenship. Agamben's conceptual work proves helpful to understanding some of the times and places discussed in this book. In the "long tunnel," for example, people were neither migrants nor refugee claimants. They were simultaneously internal and external to sovereign territory, both included and excluded, neither fully in nor out at the exceptional threshold to Canada.

In spite of his frequent textual visits to the geographical margins, however, Agamben fails to empirically ground exceptionalism—to visit the very marginal sites that he harkens. This shortcoming renders locations and processes of exclusion homogeneous and similarly homogenizes those excluded, reinforcing their sameness as an abstract concept of excluded, noncitizen outsider, or "bare life" (Sanchez 2004; Pratt 2005). But in order for state and imprisoned subject to illuminate one another, as Agamben suggests, both must be examined more closely. Although ethnography is not a method utilized by Agamben, an ethnography of the state offers this possibility by traveling to the marginal zones and administrative centers where they are created and managed.

Borders exist as walls only in the geographical imagination. In the practice of border enforcement, the border is enacted through more dispersed, chaotic geographies. Employees in the field of immigration enact dynamic networks on a daily basis. Much of their work takes place not along literal borders, but rather in office towers, across e-mail accounts and telephone wires, and among people sharing information from all over the world. Power pulsates through these networks where identities take shape, information circulates, narratives are formulated, and some groups and interests make inroads while others are excluded.

Those well positioned to tell the stories of the state are those at its bureaucratic center alongside those who have been alienated, turned out, or contained: the incarcerated, the homeless, the undocumented migrant.[8] Consider a Salvadoran asylum seeker in the United States (whose situation is documented more fully in chapter 6). After losing her husband and brother-in-law to the conflict, she fled to the United States. She holds temporary protective status to work while awaiting a decision on her case. Until that time she can work, but not borrow money. She can reside, but faces a statistically small chance of being able to stay in the country. She would like to have status, and she would like to know what the future holds, but she must wait several years to know the outcome of her case. During this time, her jobs are limited to those that will accept a work visa that is continually pending annual renewal. She cleans houses and hotel rooms and works the assembly line. She rents the smallest living space so as to maximize the remittances she sends to family at home while also proving that she will contribute to the United States if granted citizenship. She attends church, studies English, and puts money in the bank in order to prove she has a socially "settled" life characterized by economic contributions to a society now renowned for the demise of its social structures (Putnam 2000).

The state she has encountered shaped a daily life characterized by violence in El Salvador, causing her displacement to the United States. She now lives paradoxically inside of, and yet marginalized by, the very sovereign power that funded the repression of civil conflict in El Salvador. Although not being held in a federal detention center, she finds herself imprisoned by a potent blend of state violence, geopoliticized asylum policy, and bureaucratic delay. She rents, buys, eats, works, loves, and sleeps in ways that maximize her chances for asylum in the United States. She internalizes, expresses, and resists what she sees as the desires of the state all at once, paradoxically included and excluded in Agambennian terms. She grapples with the regulatory effects of U.S. asylum policy in her daily life, where the state is very much alive. In her autobiographical narrative, she offers the most powerful critique of states that are not dead in a world where borders are daily transcended by capital, corporations, and ideas—but rather alive, thriving, and themselves transnational networks. She speaks to contemporary debates happening not only in the United States, but in all immigrant- and refugee-receiving nation-states.

While theorists of transnationalism have restored the agency of migrants and decisionmaking processes to theoretical models of immigration (Rouse 1992; Silvey and Lawson 1999; McHugh 2000), the concept and the agency of the state has receded from view. And yet as John Torpey (2000, 3) argues, control of human movement is "an essential aspect of the 'state-ness' of states." Scholars are now reenvisioning the roles played by nation-states in mediating migration by scripting the identities of transnational subjects (e.g., Tyner 2000; Sharma 2001; Walton-Roberts 2001; Mountz et al. 2002). For the same reason that the complexity of migrants' quotidian lives challenged econometric push-pull models of migration (Kearney 1991; Rouse 1991; Rouse 1992; Silvey and Lawson 1999; Lawson 2000; McHugh 2000), it is equally important to include the decisionmaking processes of those involved in the governance and governmentalities of immigration. More recently, nation-states operate in insidious fashion to manipulate remote spaces where they operate outside of domestic and international law. Consider, for example, foreign-national "enemy combatants" imprisoned during U.S. military action in Afghanistan at Camp X-Ray on the U.S. military base on Guantánamo Bay. Similarly, the

United States has imprisoned asylum seekers on Guantánamo Bay and on the small Pacific islands of Guam and Tinian (United States Committee for Refugees 1999). In Giorgio Agamben's (1998) terms, these are paradoxical zones of sovereignty, at once inside and outside of sovereign power. These transnational sovereignties function by enacting sovereign powers beyond traditional sovereign boundaries. Enactments in these sites illuminate the state in all its constellations, shedding light on the practices at its center from the topography of its margins. New transnational topologies of the state can be understood only partially from the confines of sovereign territory. During times of crisis, the very geography of the state expands to inhabit and exclude, occupying new territories at its margins.

Border crossings offer striking demonstrations of the importance of visibility to states in their work on migration. Much has been made of Foucauldian conceptions of power that entail self-surveillance. As we approach the border, we know ourselves to be watched, to be more visible when we are "othered" because of race, gender, class, accent, health conditions, or personal style. But not enough has been made of the geographical context of these intimate encounters. Put simply, like human smugglers, states operate strategically in unlikely places, responding in different ways to distinct populations and migrations.

Thinking about the state as transnational actor that operates beyond traditional sovereign territory means reconceptualizing the spaces of sovereignty. As Hardt and Negri (2000) argue, state practices are increasingly diffuse. For asylum seekers, this has meant the extension of borders into international zones and the privatization of places like remote detention sites and stateless rooms in airports. In these paradoxical sites, a constellation of governmental and nongovernmental actors perform sovereignty on a day-to-day basis along a dynamic time-space continuum. The state sees and performs itself to be seen strategically. One must hear not only the voices and experiences of asylum seekers, but those of the persons who made decisions about cases and the encounters that led to those moments. In this book, the collisions of undocumented migrations, asylum seeking, human smuggling, and bureaucracy illuminate the quotidian dynamics of othering through exceptionalism in a range of geographical locations.

Tracking the State from the Inside Out

This book tells tales about "the state." These are not fictitious stories of a mysterious institution so powerful as to remain abstract and wholly detached from our everyday lives. Instead, this is a state knowable through its daily interactions with citizens and others, whether through experiences of immigration and border enforcement, or through policies that civil servants enact through their work. This book seeks to understand the quotidian practices of states through ethnography: to give up the ghostly for the mundane, the banal, the performative, and the prosaic (Painter 2006) that lead to the exclusion of migrants.

The book traces the daily life of the state geographically, beginning in the belly—the bureaucracy—and then moving outward beyond the office and into the architecture of everyday life. Each chapter tracks the state in action as civil servants board boats off the Canadian coast, run remote detention centers on islands on behalf of Australia and the United States, and make seemingly mundane decisions about asylum cases that profoundly alter the lives of people living transnationally.

The first chapter explores the global landscape of human smuggling and the challenges posed by smuggling to states in the global North. It outlines key debates among policymakers and refugee advocates regarding, for example, the relationship between smuggling industries, asylum, and enforcement policies. In asking why boat arrivals constitute crises for states, I return frequently to visibility and visuality, examining how states see and represent undocumented migrations across borders. The Canadian government's attempts to manage migration from China illustrate the creation of and reaction to crisis.

In chapter 2, I analyze the work of civil servants from the office tower where debates happen and issues arise, then on the water where these issues are negotiated "on the fly." Borders are continuously reconstituted as authorities intercept ships carrying the migrants, who are then processed on land, but not necessarily Canadian territory. "Stateness" (Painter 2006) is reproduced as migrants access legal rights and file refugee claims—though access to these rights is a struggle, not a given.

An ethnography of the state, in chapter 3, dwells inside the bureaucracy. It traces the daily struggles of civil servants charged with policing borders and exposes the *political* nature of this work. My ethnographic findings depict a performative state that excels at cultivating crisis and creating response. Crisis, in turn, creates grounds for exclusionary practices that appear exceptional by nature.

Crisis enlists corresponding geographies and discourses. The "bogus refugee" is one particular permutation of exceptionalism that links migration to exclusion in the name of national security in chapter 4. States question the legitimacy of asylum seekers, which helps to justify remote detention of refugee claimants and their subsequent disappearance from view. Legal struggles ensue as states defend—and legal communities challenge—decisions.

Meanwhile, states move offshore, farther away from the center of sovereign territory, to its very margins and beyond. Anticipatory border enforcement by states prevents migrant bodies from seeking asylum, excluding them from jurisdictions where they could make a claim. These governments extend their reach beyond sovereign territory, creating zones that render migrants stateless by geographical design. These paradoxical zones are characterized by conflicting narratives of vulnerability and control, legality and illegality, visibility and invisibility.

Finally, I map the haunting reverberations of the state through the daily lives of migrants. The state is produced and reproduced through the uncertain legal status of migrants and their daily efforts to behave in accordance with perceived expectations. The state thus emerges in quotidian form through its repercussions for migrants, asylum seekers, and citizens who struggle to negotiate policies and to survive economically. Chapter 6 explores how people living in the shadows of the state confront and enact states and their policies in their daily lives.

This tracing of the state from the inside outward suggests the need for inventive tools, methods, and political interventions through which to see state practices anew. It points to the urgency of projects that challenge the intimate exclusions of the state.

Chapter 1 **Human Smuggling and Refugee Protection**

IN JANUARY 1999 employees of Citizenship and Immigration Canada (CIC) at regional headquarters (RHQ) in Vancouver held tabletop simulation exercises to develop an operational response to a potential marine arrival of smuggled migrants off the western coast of Canada. In June, five months later, they led governmental partners through exercises on the water to practice modus operandi for a response. They rehearsed responses to hypothetical situations such as boarding unflagged foreign vessels at high speed. Immigration employees working in the fields of environmental scanning, strategic planning, and intelligence gathering recognized the possibility that one day such an event might occur.

> We were prepared. We had thought about it. We had done a contingency plan. In fact, in our plan, we had done some scenarios and workshops on the Island in the spring. One of the areas that we had actually looked at and mapped and everything was where two of the boats came in. So there had been some thinking about it. We had enough information to tell us what was happening in Australia, that it could very likely happen here, and so we were going to get ready. (Interview August 2000)

Although accusations have been made to the contrary, they did not know that at that moment smugglers were, in fact, retrofitting a fishing trawler that would transport migrants from Fujian, China, to British Columbia. Employees of CIC had devised a response with no idea of the magnitude of what lay ahead in the months to come: high-speed chases on the high seas, vessel piracy, and attempts by smugglers to undermine government procedures.

On 19 July 1999 the *Yuan Yee*, a rusted ship carrying 123 migrants smuggled from Fujian, crossed the twelve-mile boundary off the shore of

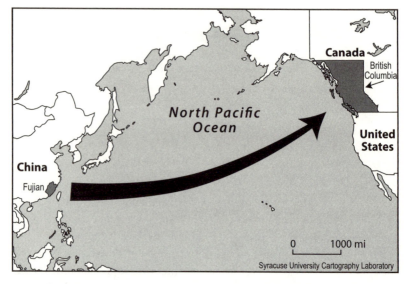

Map 1. The journey across the Pacific Ocean between the coastal provinces of Fujian, China, and British Columbia, Canada.

Vancouver Island, the western boundary of Canadian territory, and entered Nootka Sound (Maps 1 and 2).

After sighting the boat, a Coast Guard captain contacted RHQ, located in downtown Vancouver. Following is a recollection of that moment by the immigration officer who took the call.

I have to say that we didn't know in advance that we were going to get hit.... Everyone was convinced we knew, that there was some secret U.S. spy satellite up there that was telling us stuff. I took the first call on the 20th of July from the Canada Coast Guard and it was the Captain of the *Tanu*. ... I remember him saying, "You know, we've got boats." And I said, "Of course you've got boats, you're the Navy." And he said, "No, no, *you* have a boat." ... "We've got some migrants here and two of them are swimming to shore. What should we do?" I said, "Well, where are they?" He gave me the latitude and longitude, and I said, "Well, where the hell's that?" And he said, "Nootka Sound." I said, "God, way over there!" That was the first landing in British Columbia centuries ago, and here they were at Nootka Sound. I said, "Well, how far off are they?" He said, "Oh, about 100 meters from the shore." So we were going, "Wow, this is weird." We said, "What are you going to do?" And he said, "Well, what are *you* going to do?" ... And it was then we realized

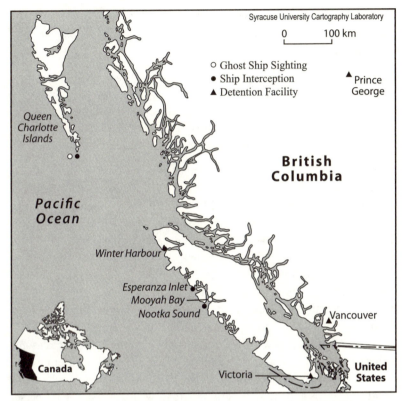

Map 2. Offshore interceptions, ghost ship sightings, and the onshore detention facilities where intercepted migrants were detained in Esquimalt, Vancouver, and Prince George.

> that we do have a federal mandate, and we felt good that we had a response that we could exercise and that we had been thinking strategically at the local level. But we didn't know until they were on the beach. (Charlton et al. 2002)

This chapter examines the daily encounters of the state with undocumented migrants. By tracing these encounters between the Canadian government and migrants smuggled from China, the chapter illustrates the attempts by nation-states to order human migration, which may be "spontaneous" and for which policies have yet to be written. The discussion begins with the global landscape of human smuggling and then addresses the phenomenon as it unfolded in Canada during one controversial and memorable episode in recent history.

The Global Landscape of Human Smuggling and Asylum

The 1951 Convention Relating to the Status of Refugees provides a formal definition of a refugee in Article 1:

> A person who is outside his/her country of nationality or habitual residence; has a well-founded fear of persecution because of his/her race, religion, nationality, membership in a particular social group or political opinion; and is unable or unwilling to avail himself/herself of the protection of that country, or to return there, for fear of persecution.

This definition was approved by a UN special conference in 1951, the same year that the United Nations High Commissioner for Refugees (UNHCR) came into being. The 1951 Convention was designed to manage European refugees in the aftermath of World War II, and the 1967 Protocol to the Convention expanded its scope geographically and temporally to help those displaced by conflicts in other parts of the world.

Migrants whose transport is facilitated by human smugglers are not necessarily refugees, but generally constitute what policymakers call "mixed flows," consisting of people on the move for any of a variety of reasons. Many migrants who employ the services of human smugglers are considered "economic migrants"; they migrate for primarily economic reasons. Others, given access to a fair refugee determination process, are found to be "Convention refugees" as outlined by the 1951 Convention—displaced for political reasons and granted status due to a "well-founded fear" of persecution upon return to the country of origin.

As states intensify enforcement practices at home and abroad, it becomes more difficult for both economic migrants *and* those displaced for political reasons to reach the relatively small group of refugee-accepting states, such as Canada and the United States, to make a claim (UNHCR Summary Position, 11 December 2000). High percentages of both groups, therefore, increasingly employ the services of human smugglers to reach sovereign territory (Koser 2000; Kyle and Koslowski 2001; Nadig 2002).

The United Nations estimates that some 4 million people employ smugglers to cross international borders annually (International Organization for Migration 1997). A total of some 200 million people live in debt to human smugglers (*New York Times* 2000a). An estimated 12 million immigrants live undocumented in the United States alone, some 3 million in Europe, and around 2.7 million in East and Southeast Asia (Skeldon 2000a, 12). Most recent estimates place global revenues from smuggling in the multibillion dollar range, which means that human smuggling has become among the most lucrative transnational businesses and now rivals drug trafficking in profitability (Kyle and Koslowski 2001, 4).

Spontaneous arrivals pit the human agency of the refugee claimant who employs determination and expends resources to migrate, against the preference of nation-states to select their own immigrants and refugees from a pool abroad. Nation-states increased border enforcement throughout the 1990s and again following the terrorist attacks in the United States on 11 September 2001, and subsequent attacks in Europe and elsewhere. They aimed to reduce the number of people migrating without authorization to work or make a refugee claim. Some states operate refugee and asylum application programs internal to sovereign territory, but they prefer to resettle people they themselves select from refugee camps abroad with the assistance of the UNHCR. Recent years have seen an incremental increase in resettlement from abroad as a percentage of overall resettlement, along with a precipitous drop in asylum seekers (see Newland 2005; BBC News 2009; United States Committee for Refugees and Immigrants 2009, 29). This signals a clear preference among policymakers to "manage" transnational migration—to exercise more control over flows rather than respond to "spontaneous arrivals," who arrive onshore unexpectedly.

As a result, and in spite of continued global conflict, the overall numbers of refugees being resettled decreased significantly throughout the 1990s, and more so following 11 September 2001 (Newland 2005; UNHCR 2007). Most dramatic has been the decline in opportunities for asylum for those on the move in search of protection and living outside of the country of origin.

Just as smugglers turn to one another to find the best routes, nation-states look to one another for policies or "best practices" in the area of border enforcement. Marek Okólski (2000) has deployed a game metaphor to explain how states see themselves in relation to one another's enforcement strategies. In this game, the world map is checkered with "stronger" and "weaker" states. Authorities working in enforcement generally believe that smugglers will exploit "weaker" states with "softer" enforcement practices through whatever methods and routes are available to them.

In the 1990s, for example, smugglers moved migrants through Europe in cargo trucks (Smith 1997, 7), while cargo boats were the modus operandi in the Americas and Australia. Over the decade the illicit movement of people in a modern-day form of indentured servitude intensified in terms of both quantity and public awareness around the globe.

Temporality is an important factor in equations of exclusion. Most immigrant-receiving states or countries, unprepared for the increased volume of applications, experienced backlogs and a long wait for the resolution of cases. This wait was seen as inhumane by asylum seekers, whereas for federal governments it became a security concern, with applicants being allowed to stay until their cases were resolved. Particularly in times of crisis, efforts to promote human rights can be at odds with national security.

In the ensuing years the interception of high-profile smuggling operations, along with terrorist attacks, fueled a xenophobic backlash against multicultural policies in Australia, England, and Canada. These incidents contributed to the scapegoating of immigrants, refugees, and immigration and refugee policies that accompanied the United States–led "war on terror" following the terrorist attacks of 2001. By December 2006 British Prime Minister Tony Blair had "formally declared Britain's multicultural experiment over" (Telegraph 2006). These ingredients completed the recipe for more securitized environments, the restricting of immigrant and refugee resettlement programs, and shrill anti-immigrant political campaigns and discourses (Bigo 2002).

Highly publicized interceptions of human smuggling operations draw attention to the conflicts nation-states encounter in their efforts to land immigrants, maintain an inexpensive labor force, build a multicultural society, and foster discourse regarding border enforcement and security in post-9/11 political environments and defense regimens. These contradictions have reached an acute level in Canada, known for its high rates of immigration, progressive refugee resettlement programs, multicultural population and policy, *and* sophisticated border enforcement practices.

In Search of Protection in Canada

Recent statistics and trends in Canada correspond with other countries around the world with major, managed refugee resettlement and immigration programs. Facing diminishing population and economic growth, Canada recruited immigrants aggressively throughout the 1990s. Such recruitment was vigorous in Vancouver, where wealthier economic migrants had been immigrating in great numbers since the mid-1980s and finding a secure environment for their savings, investments, and family members in anticipation of the handover of Hong Kong from Great Britain to China in 1997 (Mitchell 1993; Ley 2003). The federal government of Canada invested heavily in immigration as a method of counteracting a negative rate of natural population growth (Ley and Hiebert 2001) and of generating economic activity (Mitchell 1993).

The Liberal Party of Canada, in power until 2006, held high immigration rates as a long-standing policy objective and its preferred way to build a multicultural population (Ley and Hiebert 2001). Like other immigrant-receiving nation-states, Canada aims at recruiting the best, brightest, and wealthiest immigrants, but not those whose purpose is to escape poverty. As a result, economic disparities between immigrant-receiving and source countries continued to grow as wealthy business immigrants moved to British Columbia by the thousands (Ley 2003).

Canada had also for decades had the highest rate of refugee resettlement per capita in the world. Canada signed both the Convention and its Protocol in 1967. The country officially declared itself a "multicultural" country by devising multicultural policy in the late 1970s, which was passed into law with the Multicultural Act in 1988 and bureaucratized with a federal department, Heritage Canada. At the turn of the twenty-first century, Canada boasted the highest per capita rate of immigration in the world, more than two times that in the United States (*New York Times* 2002). Canada set the stage for addressing immigration and refugee issues globally, and became the place where the paradox of liberal immigration policy and contemporary border enforcement emerged most prominently.

Among immigrants to Canada from countries with refugee resettlement programs, some are sponsored by government ("government-assisted refugees") and others through private sponsorship. Those resettled from abroad can also be categorized by "source country" and "asylum country." The former are resettled directly from their home country, and the latter are resettled from yet another country where they did not fit the category of Convention refugee. In 1999 Canada hosted 53,000 refugees and asylum seekers. According to the United States Committee for Refugees, 24,732 had cases pending and 12,954 received refugee status that year. Canada resettled 9,777 from abroad, as well as 5,513 Kosovar refugees facilitated by the UNHCR humanitarian evacuation program (United States Committee for Refugees 2000a).

Although known for its leadership in progressive humanitarian efforts to protect refugees, Canada has also shown leadership in developing interception practices abroad to reduce the number of "spontaneous arrivals" to sovereign territory. In the late 1990s human smuggling by sea was on the rise around the globe, and unauthorized migrations and spontaneous arrivals were on the rise in Canada. Intelligence analysts estimated that the number of people living undocumented in Canada at that time was in the tens of thousands. More recent estimates point to some 200,000 people residing without authorization in Canada (Migration News 2006).

Canada was not alone in developing a response to human smuggling and trafficking. By the late 1990s Australia and the United States had become more aggressive in border enforcement on the water, turning away boats in the open seas before they could land.[1] The perception in Canada was that smugglers then diverted routes to other nation-states where the migrants were less likely to be intercepted or, if intercepted, more likely to be brought ashore.

Although boats of immigrants had been arriving in coastal waters now considered Canadian long before the inception of the nation-state (Guillet 1963), this was the first time in recent history that the government had intercepted a boatload of migrants smuggled over a continuous route from China to British Columbia. The arrivals linked Canada's present to its past and reminded responding officers of two other memorable arrivals in Canadian history. The *Komagata Maru* arrived in Vancouver's harbor in 1914 carrying 376 people, most of them Punjabi Sikhs (Johnston 1989). In what is now considered a dark period in the history of immigration to British Columbia, Canadian officials kept the migrants on the ship, where they stayed for two months until they were ultimately turned back. Nineteen passengers were killed when they returned to India, and most of the others were arrested and imprisoned.

More recently the *Amelie* arrived off the Atlantic coast near Halifax in 1987 carrying 174 Sikhs. This arrival caused such a stir that Parliament was recalled and a federal policy on marine arrivals drafted. This policy had expired, however, by the time the boats arrived from China.

Like these earlier ships, the boat arrivals from Fujian contributed to Canadian fears of invasion and of the possibility of a direct route and therefore continuous migration from the People's Republic of China to Canada.[2] Although the numbers were relatively insignificant compared with those who cross Canadian land borders and arrive through airports, the symbolic significance struck a chord in a nation-state historically anxious about sovereignty because of its neighbor to the south.

Like migrants running across highways or swimming across the Rio Grande on the U.S. border with Mexico, the marine arrivals gave the appearance of a lack of control and order along the border, exemplified by the political cartoon in Figure 1, which depicts migrants sailing through Canada's Immigration Act. Unlike those immigrants and refugees whose arrival federal governments facilitate, boat arrivals represent a direct affront to the integrity of national borders. They tend to evoke protective sentiments and territorial attachments. The visibility of their arrival

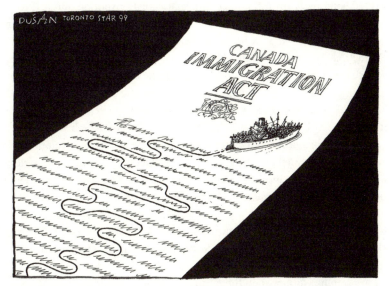

Figure 1. In this political cartoon, a ship carrying migrants sails through Canada's Immigration Act, a visual representation of criminality through evasion of the law. "Canada Immigration Act" was printed in *The Toronto Star,* August 3, 1999. Courtesy of Dušan Petričić.

provokes xenophobic reactions from the public and conspicuous enforcement responses.

Although the size of Canada's undocumented population is nowhere near that of the United States, the public response to the boat arrivals resonated with oft-repeated political reactions to undocumented migration in the United States. Canadians pride themselves in "law, order, and good governance." Yet, as Figure 1 illustrates, this migration stirred fears that the country's "generous" immigration and refugee laws were being "abused." As discussed in chapter 5, such fears, though particular in national character, are not exclusive to any receiving nation-state but are the domain of many. Canada is no exception.

A Closer Look at Canada's Place on the Map

Canada has always felt protected by its geography of sea borders and contiguity to the United States and therefore secure in its ability to select migrants and refugees rather than be selected by them. Additionally, the relatively young country has been a leader in drafting and implementing national and international policies that confront human smuggling, including the

United Nations Transnational Organized Crime Convention and its Proto-cols to Prevent, Suppress, and Punish Trafficking and Smuggling in Persons. In 2001 Parliament passed Bill C-11, the Immigration and Refugee Protection Act, which expanded the government's capacity for front-end security and detention. Canada also signed the Safe Third Country Agreement (STCA) with the United States as part of the Smart Border Accord. The STCA restricted access to asylum in Canada by preventing those entering Canada by land from the United States from making a refugee claim.

Yet Canada's geographical proximity and otherwise close relationship with its southern neighbor has also produced a distinct climate of anxiety. Canadians have always feared an erosion of sovereignty, and more so in a neoliberal age of free trade and economic integration where cross-border disputes often erupt about everything from the trade in softwood lumber to smuggling along the forty-ninth parallel. Scholars studying the border shared by Canada and the United States have found that while Americans tend to prioritize national security over other concerns, national sovereignty remains foremost on the minds of Canadians (Nichol 2005).

With nation-states such as the United States, Australia, and the United Kingdom increasing enforcement measures, Canada is under pressure to do the same or face increased traffic in human smuggling. This is precisely the state of affairs of the mid- to late 1990s. Refuge-granting, immigrant-receiving states were restricting asylum policies and stepping up enforcement.

On 6 June 1993 a cargo steamer named the *Golden Venture* entered New York Harbor and ran aground near Rockaway Beach. Some 286 people from Fujian Province in China who had endured a perilous 112-day journey jumped into the water. Most were intercepted by the authorities. Ten drowned in their flight, and the U.S. media and spectators became engrossed in the arrival. Peter Kwong (1997, 3) characterizes the passengers as "less a threat to American borders than survivors of a misdirected ship of fools that had unwittingly crashed right into the financial and media center of the world."

Nonetheless, following the grounding of the *Golden Venture*, the United States government designated an interdepartmental task force to police waters and aggressively combat human smuggling, to reduce the number of Chinese migrants traveling by boat. Smugglers are believed to have then moved clients via other geographical routes, through Central America, Mexico, and the Caribbean (Smith 1997, 4; see DeStefano 1997). Canada's role as transit country and "problem spot" also became more pronounced

(Smith 1997; Yates 1997). Civil servants tasked with policing were aware of the scope of the problem. One CIC employee said, "We can afford to be a bit smug, but it's not because of anything we're doing right, but because of geography and an aggressive neighbor" (Interview, Victoria, March 2001). In the years prior to the marine arrivals, Canada had been a more desirable transit country for the industry moving migrants to the United States (Yates 1997), which subsequently began pressuring Canada to step up enforcement along its borders. The United States has long called on Canada to strengthen enforcement and harmonize its policies with theirs, sponsoring conferences on harmonization and integration that mirrored EU regionalization. This pressure grew more acute as smuggling by sea increased in the 1990s.

In the search for answers following the September 11 terrorist attacks, much concern in the United States was directed at Canada's border controls and refugee policies. One immigration employee said, "It's cliché to say it, but the world is smaller, organized crime is very proactive, and we're reactive. By the time we do anything, they are already a part of Canadian crime. Criminals look for the soft underbelly, and they're finding it here in Canada" (Interview, Victoria, March 2001). Civil servants on both sides of the border suspected what their publics feared: that the undocumented population of migrants moving to and through Canada was on the rise.

Prime Minister Pierre Trudeau once remarked to his nation's southern neighbor, "Living next to you is in some ways like sleeping with an elephant. No matter how friendly and even-tempered is the beast . . . one is affected by every twitch and grunt." As far as Canada is concerned, the United States is never distant. Could the world's humanitarian diplomat really neighbor the sleeping elephant to its south and not take a turn toward stricter enforcement? This has been a source of continuous debate among Canadian civil servants and civil society, and a concern consistent with what Emily Gilbert calls a "discourse of inevitability" along the border (2005, 203) that posits regional integration between these neighbors as inevitable.

The extent to which the United States influences Canadian public policy is a matter of debate, but the perceived pressure has been significant for civil servants. Widely publicized enforcement cases feed the public discourse and the fears of public servants. In late 1999 one significant incident took place at the border between British Columbia and Washington State when Ahmed Ressam was intercepted crossing from Canada into the United States, allegedly planning to attack LAX Airport in Los Angeles.

Ressam was a refugee claimant from Algeria who had been living for several years in Montreal. Although his refugee claim was rejected and a warrant issued for his arrest, he managed to evade authorities until his arrest at the crossing between Canada and the United States at Port Angeles, Washington, in December 1999, when materials for a chemical bomb were found in his car (*National Post* 2001).

Ressam's interception planted the seeds of an intense fear of dangerous immigrants crossing over from Canada that followed the terrorist attacks on New York and Washington, D.C., on 11 September 2001. Amid the flurry of speculation, North Americans were led to believe that some of the 9/11 perpetrators might have entered the United States from Canada. The angst was misdirected: none of the hijackers involved in the attacks on September 11 had migrated to Canada as immigrants or refugees, nor had any crossed into the United States from Canada as U.S. government officials had initially believed. Regardless, the forty-ninth parallel began to occupy a more prominent place in the American geographical imagination. In spite of the fact that more people crossed north from the United States into Canada to make refugee claims than traveled in the reverse direction, Americans came to perceive Canadian immigration and refugee policies as a threat.[3]

Following the interception of Ressam, the perception of hidden geographies behind the terrorist attacks intensified the public discourse, with accusations of links between Canada's refugee program and "illicit" movement through North America. The parallels between the discourse on the "war on terror" and the strategies to combat smuggling are striking. Both phenomena involved networks that transcended international borders. Both operated in transnational ways that challenged states in their responses, and both threatened public support for refugee programs. Both involved efforts on the part of nation-states to combat a faceless, nameless set of individuals and networks based in multiple states. This invisibility, coupled with racialized fears of terror and violence, resulted in the criminalization of migrations and in receiving states pressuring those regimes that hosted terrorism or smuggling in a global game of diplomacy. Both trends demanded improvements in intelligence-sharing capacity within and among states, as well as other strategies to curb smuggling and terrorism, such as freezing the assets of terrorists and smugglers and following the money trail. Both posed challenges to overtures toward more open borders and border policy harmonization between the United States and Canada. Both applied more pressure on Canada to increase enforcement. And finally, neither human smuggling nor terrorism fell neatly into the mandate of any one federal department in either Canada or the United States.

The ambiguous geographies associated with human smuggling often give rise to the circulation of incorrect information, the conflation of refugees with terrorists, and the scapegoating of immigrants (Honig 1998). These migrations fanned long-standing tensions between Canada and the United States regarding border enforcement and refugee claimant policies and resulted in changes to policy that intensified not only border enforcement, but the refugee claimant process itself by increasing the frequency of front-end security checks, for example.

Western receiving countries such as Canada and the United States often believe incorrectly that they can turn streams of transnational migrants on or off as they would an electric switch. But shifts in public sentiment and political will toward immigration in the receiving country do not necessarily correspond with the conditions that contribute to or compel out-migration in the source country. Complicating the smuggling environment in Canada is its geographical proximity to the United States. Substantial evidence shows that the migrants on the four boats from Fujian Province intended to join the large undocumented population in the United States but made refugee claims in Canada when intercepted by Canadian authorities. As it pressured Canada and Mexico to enforce their border security and curb migration streams, efforts that culminated in the signing of the Smart Border Declaration in December 2001, the United States helped contribute to the growth in undocumented migration because of the significant role that migrant labor played in supporting the booming U.S. economy in the 1990s. The steady demand for cheap labor along with the increased militarization of border enforcement fed the human smuggling industry.

Human Smuggling by Sea from China

In many ways, there was nothing unusual about the arrival of boats of immigrants to North American shores. The history of the settlement and growth of Canada and the United States is characterized by the arrival of newcomers by sea. Moreover, China's coastal history is characterized by centuries of out-migration from Fujian (Hood 1997; Kwong 1997; Chin 1999), although the channels and routes change perpetually (Skeldon 2000a). According to Ko Lin Chin (1999, 37), the contemporary practice of boats smuggling Chinese migrants to North America has been going on since 1989, when "modern smuggling by sea began." Countless human tragedies occurred during these episodes, some documented—such as the deaths of fifty-eight Fujianese migrants in the back of a refrigerated to-mato truck in Dover, England, in June 2000—and others not. The great tragedy behind the gaps in empirical knowledge of human smuggling is the disappearance of migrants crossing deadly terrain, whose fates can only be imagined (Moorehead 2005; Nevins 2008). The smugglers' shift to un-flagged cargo ships was first recognized in North America in June 1993 with the grounding of the *Golden Venture* in New York Harbor (Kwong 1997; Chin 1999), prompting Kyle and Koslowski to label 1993 "the year human smuggling crashed into our living rooms." A parallel can be drawn here with the 1999 marine arrivals in Canada, that being the year when human smuggling "crashed into the living rooms" of Canadians.

Such high-profile episodes called the world's attention to human smug-gling practices, many involving Chinese migrants. Some of the most brutal smuggling experiences uncovered during the last decade (Smith 1997, 11) and some of the most extensive, sophisticated, and highly publicized smug-gling movements have originated in Fujian Province (Salt and Stein 1997, 474; Skeldon 2000b).

Using the paradigm of Okólski's (2000) "global game board," the expe-rience with human smuggling by Canadian civil servants is one of disad-vantaged players. Immigration officials working in the field of enforcement frequently have mentioned that they feel as if they are operating in the "soft underbelly" of the United States. This orientation resulted in a conspiracy theory expressed by a few respondents in interviews that followed the 1999 interceptions. Migrants from one boat mentioned being stopped at sea by a large, red, official-looking boat with "dark-skinned" people on board. CIC employees who shared this story interpreted this to mean that the migrant ship had been stopped on the water by the U.S. Coast Guard. The story con-tinues that this official boat refueled the migrant ship and pointed it in the direction of British Columbia. Some believe that the United States inten-tionally let these ships slip through their heavy surveillance of the Pacific coastal waters to send a message to Canada to improve enforcement and take on its share of policing along the coastline. Although the veracity of these claims has not been established, it is useful to explore their mean-ing. Canada sees smuggling from a unique perspective, perched in a loca-tion where the United States figures prominently. As in many other policy arenas, Canadian authorities often look southward over their shoulders as they devise strategies to combat smuggling.

With the marine arrivals from China, conflicts over agendas within the federal government became more pronounced. As it continued to recruit skilled immigrants, landing a total of 189,911 immigrants in 1999 (Citizenship and Immigration Canada 2000, 3), CIC faced criticism for failing to maintain its enforcement mandate. The tension between these objectives—to land approximately 1 percent of the population annually (Ley and Hiebert 2001) and to police borders effectively—is a backdrop for the daily work of federal employees in immigration as well as other civil servants. Human smuggling rarely falls neatly under the jurisdiction of one federal agency. In fact, Canada had ten federal departments involved in its response to the problem, and many challenges in the federal response involved struggles over the differing mandates among departments.

The complexity of the response began on the water. Under the leadership of CIC, various departments scrambled to authorize the deployment of the Coast Guard's *MV Tanu*, the platform from which members of CIC's Marine Response Team, the RCMP Emergency Response Team, and a medical team of three would board the unmarked vessel. As CIC intercepted and processed the migrants, little did the department know that a second boat of migrants was en route to Vancouver, with a third and a fourth preparing to leave China. The institutions that would become involved in the ensuing months anticipated neither the tone of the public response to human smuggling by boat nor the ways in which the pitch would intensify with each of the subsequent three arrivals—a total of 599 migrants coming to Canada, with the fourth and final ship intercepted on 9 September 1999. This came to be known among local bureaucrats as "the Summer of the Boats," encompassing the largest group of unanticipated refugee claimants CIC had received in recent history, and this new scale demanded an investment of resources on the parts of several federal, provincial, and nongovernmental bodies far exceeding their annual operating budgets.

After intercepting migrants on the water, authorities brought them either by ship or by bus to a Canadian military base near Victoria on Vancouver Island. Migrants were processed, including initial interviews with immigration officers. During this time well over 500 of the 599 migrants made refugee claims.[4] After being released, many that arrived on the first boat failed to appear at hearings in subsequent weeks, thus abandoning their claims. They were presumed to have traveled to New York's Chinatown to work, which made Canada a transit country in this migration. This abandonment further inflamed public opinion regarding the perceived abuse of Canada's refugee program and the policing of borders. It also fueled an increasingly securitized response from the federal government.

With the ensuing boat arrivals, the federal government argued success-fully that because the migrants were without identification documents they constituted a "flight risk"—meaning that they were likely to flee if released—and were therefore subject to detention. This marked the first time in recent Canadian history that the federal government detained refu-gee claimants en masse as they saw their claims through the process.[5]

Canada detains refugees much less frequently than Australia, France, or the United States. The Immigration and Refugee Protection Act was passed in 2001, allowing more extensive use of detention. After it was imple-mented in 2002, however, Canada still had an average of some 450 persons detained at any time (Gavreau and Williams 2003). This relatively small number illustrates the magnified significance of the refugee claimants from Fujian in the national context.[6] Eventually, the Fujianese migrants would represent the largest mass deportation in recent Canadian history.

Following the processing of the ensuing arrivals, CIC either released migrants or transported them to longer-term detention facilities in the inte-rior of British Columbia. Ultimately, of the 599 migrants, 429 (72 percent) were held in long-term detention.

CIC also placed about one hundred minors in the custody of the Ministry of Children and Family Development of the province of British Columbia, deemed the legal custodian for unaccompanied minors. The ministry cre-ated a Migrant Resource Team and quickly developed one of the most extensive programs in the world to support unaccompanied minors.[7]

According to the decision in the 1985 *Singh v. Canada* case, the right to make a refugee claim is considered a human right in Canada protected by the Canadian Charter of Rights and Freedoms. The federal government granted the claimants due process under the Charter. In some cases, detention lasted up to two years (DAARE 2001, 5) as claimants exhausted opportunities for due process. Most were ultimately denied refugee status by the Immigration and Refugee Board (IRB) according to the 1951 UN Convention Relating to the Status of Refugees and ultimately repatriated to the People's Republic of China.

The government responded to the boat arrivals employing an enforce-ment framework while fulfilling its obligations as a signatory of the 1951 Convention and adhering to the Canadian Charter of Rights and Free-doms. The smuggled migrants were generally not categorized by the gov-ernment as either the immigrants that Canada worked tirelessly to recruit or Convention refugees determined to have a well-founded fear of return. This uncertain status influenced new immigration legislation proposed in the spring of 2000, months after the first boat arrived, which was implemented in 2002 as the Immigration and Refugee Protection Act. Even though this legislation was under way well before the boats arrived, then minister of immigration Elinor Caplan presented Bill C-11 as an attempt to "close the back door" to irregular migration in order to "open the front door" to legal immigration (CIC news release, 14 June 2000). "By saying 'No' more quickly to people who would abuse our rules, we are able to say 'Yes' more often to the immigrants and refugees Canada will need to grow and prosper in the years ahead," said Minister Caplan (CIC news release, 21 February 2001). Bill C-11 was revised and resubmitted in 2001. The legislation introduced heavy penalties for human smuggling, including $1 million and up to life in prison.

Although human smuggling occurs across Canadian land borders and through airports in greater numbers than by sea on a day-to-day basis, the boat arrivals dominated the attention of the Canadian media. Photographs such as the one in Figure 2 showing rusty cargo ships with migrants crowding the deck were accompanied by unusually large headlines. "ENOUGH ALREADY" screamed the front page of *The Province* on September 1, 1999, suggesting that the tolerated number of people entering Canada had been surpassed.

The media portrayed the marine arrivals as a crisis for the federal government. The mounting dramatization of the events sparked anti-immigration sentiment to a degree Canada had not seen in many years (McGuinness 2001; Hier and Greenberg 2002; Mahtani and Mountz 2002). British Columbia's economy had been waning since 1996 and was destabilized further by the Asian economic crisis of 1997–1998. In terms of investment, the wealthy participants in the Hong Kong migration had not provided the miraculous answer to the province's economic woes for which many had hoped (Ley 2003). Meanwhile, the Canadian Alliance, a right-wing party interested in lowering rates of immigration to Canada, was gaining power across the country, as was the conservative BC Liberal Party in the province (which would assume leadership there in 2001).

The migrants who arrived by boat had not planned to stay in Canada, and neither Canada's economic woes nor its political climate were of much interest to the smugglers overseeing their journey. Once they were intercepted off the coast, however, most made refugee claims and entered a long, bureaucratic claimant process for whose thoroughness Canada is praised worldwide. Within this processing time lapse, nonetheless, more conservative politicians throughout North America accused the Canadian government of harboring terrorists and other criminal elements, whom they

A rusting freighter packed with Chinese migrants is escorted by the Canadian Coast Guard vessel Tanu near the coast of Vancouver Island yesterday.

Figure 2. The third migrant ship intercepted off the coast of British Columbia. The headline conveys the perception in the local media that Canadians were growing intolerant of migrants arriving by boat from China. Copyright *The Province*.

charged with "abusing" a "generous" system that granted them temporary refuge until the outcome of their case was determined by the IRB.

Like many "native-born" nationals living where refugee resettlement takes place, Canadians often harbor anxieties (cf. Hage 1998) about the erosion of national borders, which are exacerbated by the popular perception that migrants have abused a generous refugee determination process. Figure 3 is a political cartoon from the *Victoria Times Columnist*, a newspaper based in Victoria. Victoria is the capital of British Columbia, located on the southern coast of

Figure 3. In this political cartoon, migrants are portrayed as lackadaisical and well-dressed as they disembark, tipping the Canadian authorities, whom they expect to carry their bags. The authorities look meek and powerless, with mouths closed and eyes averted. Copyright Adrian Raeside; originally published in the *Victoria Times Colonist*.

Vancouver Island, close to the border with the US and Esquimalt, where claimants from the boats were processed. The cartoon shows migrants disembarking from a boat and cavalierly tossing a tip to enforcement officers, asking them to pick up their bags. The migrants are depicted as arrogant and happy, while the Canadian authorities appear to be immobilized, helpless, and frustrated.

This portrayal reflects a debate among scholars regarding the celebrated dissolution of political borders for transnational migrants. Early theorists of transnational migration (e.g., Kearney 1991; Rouse 1991) who defined "transmigrants" as people who "develop and maintain multiple relations—familial, economic, social, organizational, religious, and political—that span borders" (Glick-Schiller et al. 1992, ix) failed to account for the constraints imposed by borders on refugees and asylum seekers (Bailey et al. 2002). This view of economic migration suggests that human smugglers take advantage of easy passage but overlooks the reality that many migrants in search of protection employ smugglers. These mixed flows pose significant challenges for states in a politicized climate when concerns over national security and border enforcement come to the fore.

"Saving Face" in Times of Crisis: Refuge, Refusal, or "Policy on the Fly"?

The phenomenon of human smuggling has engendered a series of important debates that go to the heart of the meaning of immigration and refugee protection in the twenty-first century. Who has a right to move, for what reasons, and what kind of protection can they expect from states? What kind of access to refugee claimant processes must states ensure? In what ways can states control this access without violating their responsibilities as signatories to the Convention? Will states continue to balance mandates to facilitate labor migration, police borders, and resettle refugees? Will they continue to maintain the responsibility of protecting spontaneous arrivals, or do their enforcement actions abroad preclude protection of this particular population?

It is difficult for the international community to come to an agreement or collaborate on this problem. Contradictory evidence supports different arguments about the relationship between human smuggling and refugee determination processes. The international community of states, intergovernmental entities, and suprastate bodies attempting to regulate human migration have yet to agree on the nature of this relationship and to harmonize policies. Refugee advocates argue that increased front-end control will prevent Convention refugees from reaching a place like Canada to make a refugee claim, pushing them into the hands of smugglers. Some researchers argue that immigration is on the rise overall, with human smugglers simply satisfying a greater demand, while others argue that tighter immigration controls have in fact spurred the increased employment of human smugglers (Koser 2000; Kyle and Dale 2001; Nadig 2002). Opposing positions have become more entrenched with the dearth of empirical evidence. More research needs to be carried out exploring this relationship.

Furthermore, it appears to be in the short-term interest of states to refrain from establishing set policies regarding these issues and instead opt for ad hoc arrangements. Canadian civil servants involved in the British Columbia interceptions have called this "policy on the fly." Australians refer to such arrangements as "policy on the run." Following highly publicized interceptions of human smugglers at sea, countries often resort to what some refer to as "shopping," where they make bilateral or multilateral arrangements to either return refugee claimants or send them to a third country that steps in to diffuse a heated situation.[8] This practice allows potential host countries to "save face" on border enforcement practices through diplomatic arrangements without denying claimants an opportunity for protection elsewhere. The United States and Australia are especially fond of this option, and in April 2007 the two countries actually announced an informal arrangement to "trade" some two hundred refugees (McGuirk 2007).

Anxious encounters between states and migrants increase the prominence and magnify the role of international borders wherever they occur, whether on land or at sea. They demarcate the margins of the nation-state where power differentials are produced amid moments of crisis. The question then becomes how to conceptualize the state that performs so many functions relating to transnational migration.

This chapter set the stage for analyzing the heated response to the boat arrivals in Canada in 1999. It also placed Canada in a more global context, a landscape where people in search of protection employ the same human smugglers and pass through the same refugee claimant system as labor migrants.

The response to the boat arrivals offers a window into many of the debates surrounding border enforcement and security issues that intensified following September 11, when the discourses surrounding refugee policy and terrorism converged. Canada is an example of, not an exception to, these debates. Boat arrivals from Europe, South Asia, and China have played a central role in Canadian history. As noted in the vignette that opened this chapter, regional bureaucrats had been preparing for marine arrivals for months. Why, then, did the arrivals become a crisis within the bureaucracy, in the media, and in public discourse? To answer this question, we must consider issues of vision: how states themselves see human smuggling and how human smuggling is made visible to the public.

Chapter 2   Seeing Borders Like a State

THIS CHAPTER ENGAGES principles of vision and visual registers to aid in understanding how states see borders and how they deploy visuality as an affective register through which sovereignty is secured (see Amoore 2007).[1] The state sets its sights on transnational migrations, and ultimately becomes transnational by enacting enforcement practices along borders. Visibility proves crucial to understanding how states respond to migrants and how they catalyze publics to respond in ways that enable the advancement of enforcement agendas during highly publicized, visible, visual, and seemingly exceptional crises along their borders.

When I first visited RHQ in Vancouver in the fall of 1999, a large computer-generated map of British Columbia hung on the wall near Central Command. The geography was overwhelming those responsible for policing Canada's borders. Civil servants in British Columbia had repeated on many occasions that geography was a fundamental concern that the people at National Headquarters in Ottawa could not appreciate. The person showing me the map waved a hand above him to show "where we've been hit." This simple gesture expressed the feelings of many civil servants working in the dynamic fields of human smuggling and border enforcement. He conveyed a sense of the public's vulnerability in remarking on the length of the coast: some 900 miles from the continental U.S. border north to the boundary with Alaska and the Yukon (Map 1). Red pins on the map demarcated the locations of the interceptions as well as that of a "ghost ship" believed to have gone undetected in early July 1999. The setup was reminiscent of the popular game "Battleship." Rather than ships being hit by people, however, it was the people in this office who felt—often using the very word—they were being "hit" by the ships. In fact, "when we got hit" was the opening phrase to many stories infused with militaristic language.

Like states engaging in transnational practices, human smugglers are fluid transnational actors that have proven their ability to undermine state regulation by exploiting the division between foreign and domestic spaces. Other scholars have demonstrated how states and smugglers compete (e.g., Salt and Stein 1997; Heyman 1999; Andreas 2000; Okólski 2000; Spencer 2000). Smugglers and their clients move across nation-states and work the spaces between them. The transnational nature of smuggling limits the ability of nation-states to control mobility. This chapter addresses the variety of challenges human smugglers pose to states, and the variety of strategies and policies that states have pursued to suppress undocumented migration. It juxtaposes the relative mobility of smugglers with that of civil servants, and the relative visibility of intercepted migrants with the relative invisibility of those who intercept. Key to understanding the dance between smugglers and states is geography and visuality: how each side sees the landscape, and how each plays on distance and proximity to its advantage.

Drawing on James Scott's (1998) approach, it is important to see human smuggling from the perspective of a state to understand the disparities and relationships between human smugglers, nation-states, and the global communities in which they act. Scott uses the metaphor of vision to explore the spatial relationships between the administrative centers of states and the policies they attempt to implement. Invariably, the regional execution of state projects on the ground brings unforeseen consequences. Scott characterizes the differences between intentions and effects as the outcome of distance. Vision and linguistic exercises in classification are central to understanding the perspective of the state. As Scott argues, states perpetually attempt to impose order on chaotic realities that far exceed the state's capacity for ordering. States thus meet with varying degrees of success and resistance in different locations.

The map in Figure 4, for example, was one of several images included on a poster circulated by the U.S. government to citizens and civil servants in Canada. The poster was designed and circulated to inform and enlist others in the recognition of routes and methods by which smugglers were entering and transiting North America.

Scott does not deal with the daily practice of "seeing" like a state, yet his ideas can be applied to the day-to-day implementation of policy. In their programs to receive and process migrants, state bureaucracies must categorize distinct transnational flows. They must do so from a distance characterized not only by geography but by bureaucracy and social location. Administrative management strategies often inhibit a nuanced understanding of human migration. Civil servants face constraints imposed by international borders,

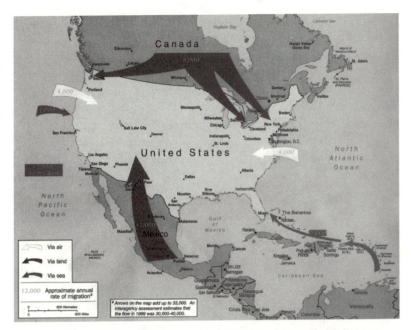

Figure 4. This map, circulated by the U.S. Immigration and Naturalization Service in 2000, shows fear of undocumented migration to the United States from Canada and other areas, represented by arrows showing numerical estimates of unauthorized migrations in 1999.

law, financial concerns, and a multitude of jurisdictional boundaries. As a result, they behave reactively in response to human smuggling, employing practices such as interception, detention, and prosecution. Smugglers, contrarily, work strategically at a more intimate local scale precisely to elude the bureaucratic vision of a nation-state that imposes grids, borders, and order upon the landscape from a distance (Scott 1998). Not bogged down by bureaucracy, smugglers can alter their smuggling routes and methods more quickly than nation-states can adapt their investigative or preventative strategies. In other words, it is easier for smugglers to elude the law than it is for governments to enforce it. Smugglers appear to be better equipped to employ flexibility in relation to the local geographies of international borders and global migrations than are the corresponding authorities of nation-states.

States see landscapes from a distance (Scott 1998) and attempt to impose order on people and places that are not subject to that order. The same can be said of the immigration and refugee policies through which states attempt to regulate transnational migration. Inevitably, people do not all fit easily into one category or another, such as "economic migrant" versus "political refugee." People tend to envision their own identities not in the terms set by states but rather as occupying and migrating through the gray areas between categories, policies, and states. Nonetheless, important struggles take place over the categories and language of migration, which have great material and legal consequences in terms of migrants' opportunities to stay. By the time migrants arrive on sovereign territory, states have already begun to define in their own terms who they are.

For Louise Amoore, visuality occupies a central role "on the home front" of the war on terror. She argues that "the emerging watchful politics is vigilant: it 'looks out' with an anticipatory gaze" (2007, 216), and that this visuality becomes the primary affective sense mobilized by the state as sovereign, "the sense that secures the state's claim to sovereignty and legitimates violence on its behalf" (217). The remainder of this chapter follows Amoore's understanding of vision and visuality as modes through which representation "categorizes and classifies people into images and imaginaries of many kinds" (217).

State Imaginations

During the year following the arrivals to British Columbia, two CIC employees traveled up the coast and boarded helicopters to photograph a series of towns where they believed off-loading could occur during future operations.[2] They then compiled photo albums presenting aerial images of the Pacific coast. These photo albums show remote locations that few Vancouver inhabitants have ever seen but that are familiar to both civil servants and human smugglers. The photographs were used not only to document interceptions, but also in predicting future staging areas for smugglers and planning potential governmental responses.

These photographs illustrate the powerlessness felt by those responsible for ensuring the safety and success of interceptions in British Columbia. Civil servants were charged with documenting the length of the coast, its geographic isolation, and the challenges facing a possible response in terms of the natural and built environment, including rugged terrain, the availability of infrastructural resources such as helicopter pads and hospitals, cold weather, and rough waters. Ironically, despite any sense of control over geography the aerial photographs provided, as explained by the bureaucrats who took them, these images also convey the overwhelming challenges posed by the geography of the coast and its defense against smuggling.

The boats that arrived some months later certainly matched the description and expectations of those who had been trained to board them earlier in 1999. Photographs taken by the boarding intelligence officers depicted the physical qualities of the boats, migrants, and smugglers. They provide insights into the characteristics sought, categorized, and documented by the emergency responders who boarded, forming the basis of a visual narrative of the interception.

The Approach

The first photographs capture a truly postmodern moment. They were taken by one of the first CIC employees flown to the site of interception from downtown Vancouver. A local television newscaster reported on the boat arrival and showed a map of its approximate location. The second image showed a yellow chopper that had landed, thus delaying a second, red chopper from setting down. The first chopper carried members of the media and the second chopper carried CIC employees. The media not only arrived first on the scene, but delayed the first authorities' reaching the site in so doing. These images reflect the photographer's awareness of the magnitude of what was happening and of the role and active participation of the media—even in a remote location along a rugged coastline—from the moment the federal response began. They reveal how quickly reporters were apprised of the interceptions and foreshadow the role media coverage would play in fomenting the ensuing crisis.

Subsequent photographs bring the interceptions into focus, showing boats that look barely seaworthy carrying migrants below deck. Many of these shots also show authorities on the water waiting to board from a variety of platforms, from the large *MV Tanu* of the Coast Guard to Zodiac (inflatable) boats belonging to the RCMP and other smaller ships, including pleasure craft with spectators and journalists on board. There, authorities watch initial communications to the boats' captains to stop and self-identify. These shots show members of the Marine Response Team (MRT), each in full white uniform with hood, awaiting their turn to board. At least one of the boats resisted interception for a time, and the federal government executed flybys to scare them into submission.

Another photograph shows authorities looking on from a ship in the foreground at the interception of the fourth boat, listing dramatically as a plane does a close flyover to scare the captain into conceding capture. Another photograph shot during these moments of encounter shows a white T-shirt flying on the flagpole on the deck of the second ship as a sign of surrender.

The Encounter

The shots then transition to the actual approach of the MRT and Emergency Response Team (ERT) via Zodiac, followed by their boarding the boats, weapons drawn. These images bring viewers into the interception, enabling them to see the ship for the first time from the perspective of the civil servant. The listing of ships, the movement of water, and the twelve-foot swells give a sense of the action in the episode that made the transfer from large ships to the migrant ships via Zodiac challenging.

On the Water

The CIC operation to find and intercept ships was named "Operation Osprey" after the tendency of ospreys to hover above the water and then swoop down quickly on their prey. When immigration officers embarked on Operation Osprey for the first time in July 1999, their dependence on other federal departments proved an immediate source of stress. CIC held ultimate responsibility for responding to a ship of migrants but had none of the resources to do so. As one manager explained, "Smuggling is an immigration problem, but we have no assets; just human resources" (Interview, Victoria, March 2001). A successful operation on the water therefore required working relationships with "federal partners," such as the Canadian Coast Guard and the Department of National Defence, which supported the response with surveillance planes, vessels that served as platforms from which to board the boats, and a site for processing.

Many different governmental departments were involved. Some, such as the RCMP, operated within an investigative framework to suppress and prosecute smugglers; others, such as Status of Women Canada, worked from a humanitarian perspective to provide access to asylum and to potentially "protect" people from human trafficking (see Sharma 2005). The federal departments involved in the response had differing mandates and distinctive institutional cultures that related particularly to their status as investigative bodies. CIC is often seen as "soft" from these departments' investigative and enforcement perspectives and unable to protect information. In the early 1990s, CIC lost its status as an investigative body, and with it the right to access and the capacity to protect certain kinds of information. There was tension among the various institutions involved concerning the sharing of information and about goals in an operation and subsequent investigation. While the RCMP had primarily investigative concerns, for example, CIC employees worked to enact mandates to both enforce border control *and* protect refugee claimants.

These institutional differences manifested in the day-to-day interactions among federal employees and factored into the collaboration during tabletop exercises held in the spring and again during the actual marine responses in the summer. The emergency response teams of CIC and the RCMP overcame these tensions in their work on the water through effective leadership. One operational leader on the water explained how lucky he believed the department was to have him involved. As a former "military man," he understood the formal hierarchy underscoring military operations and "what the patches on officers' sleeves meant," even though CIC did not follow such a hierarchy. He also explained that because he himself was a man with no large ego to massage, he was able to interact well with managers in other local offices to finesse the teamwork and trust required among agencies for a successful interception (Interview, Victoria, March 2001).

There were other challenges to overcome, however. One conflict involved the demarcation of national territorial waters versus the "high seas" or international waters. The arms of the state could not agree on its basic geography. This boundary lies twelve nautical miles offshore and marks the beginning of the jurisdiction in which Canadian vessels can legally approach suspicious vessels. The boundaries of interception and arrest were negotiated not only on the water, but also via satellite telephone between those leading operations on the water, those coordinating from Central Command in Vancouver, and those following and dictating the action from national headquarters in Ottawa. Civil servants responding on the water complained that lawyers in Ottawa office towers delineating the legal edges of sovereign territory for the purposes of interception did not understand the geographical challenges to responding within a given range on rough seas along a rugged coast. Their ignorance of geography only served to reinforce Scott's assertion that projects designed to order the landscape, conceived at the administrative center, tend to fail.

The interception of the second boat involved a particularly harrowing set of circumstances for both the migrants and the responding authorities. Once the boat had been sighted, a chase ensued on and off for nearly two days. During this time the crew of the ship managed to drop off the migrants—including the largest group of minors to arrive among the four boats—and ordered them to hike over a rise to a village to meet contacts once they had reached the rocky shore. No such village existed in this remote area. When authorities arrived, the migrants had started a fire but were suffering from hypothermia because they were wet and inappropriately dressed for cold temperatures. Meanwhile, the cargo ship was fleeing for international waters at a high speed.

While CIC's MRT performed medical triage on the beach and search and rescue in the surrounding area, an effort complicated by the inaccessibility of the beach (authorities discussed the possibility of having supplies air-dropped at a location where they could set up camp for a few days) (Interview, Victoria, March 2001), the RCMP went after the ship that had dropped the migrants and was now on the run. A hot pursuit ensued, along with several miscommunications. The boat evaded capture, forcing the Coast Guard to pursue at high speeds in the *MV Tanu* in a cat-and-mouse chase. With the media reporting the ship's arrival, CIC was already in the limelight and being pressured by Ottawa not to miss this interception. At one point during the pursuit the boat disappeared from the horizon, concealed by a quick succession of erratic changes in direction and a thick fog. One respondent described a moment of silence on the accompanying conference call during which participants on the line wondered who among them would lose their jobs.

When authorities finally approached the ship, the crew would not respond to commands shouted in Mandarin to stop and identify themselves. As it turned out, the nine crew members spoke Korean and did not understand Mandarin.[3]

On the Ship

Once on the ship, the photographer sought clues to explain what had taken place. Photographs included shots of the ships themselves, of objects found within, and of people. The close-up shots of the ships detail their condition, the rusted exterior, and particular aspects of the decks, one of which was mostly covered by long tarps.

The objects documented provide clues to the lives and hopes of those on board. The photographers document how the ships were retrofitted to make these journeys. Investigators found U.S. bills stashed in women's sanitary napkins, along with letters, journal entries, and navigational charts. Another image shows the blue-gloved hand of an investigator pulling maps out of a drawer on the bridge.

Investigators and analysts expressed interest in the machinery that supported the trip, such as generators, and the technologies used by the crew, including cell phones, radar, and GPS units, which had often been jettisoned by the time of interception. These clues tell them about the destination, but do not help them identify the people, most of whom did not carry identification.

Photographs of people show the MRT grouping migrants by sex and by role; those seen walking around freely on deck were identified prior to boarding as suspected enforcers.[4] Photos following the boarding show such enforcers tagged, handcuffed, and separated from their clients on deck.

On the ships, the photos portray authorities trying to impose order and administer to health and security needs. In the very cold coastal air, migrants sit on decks huddled under blue blankets and drinking bottled water distributed to them by authorities. Some look frightened and anxious; others appear relieved. Suspected enforcers, on the other hand, are shown wearing plastic handcuffs, some maintaining a defiant countenance. The migrants are usually grouped on deck, although one image shows them crowded and seated in the hold below deck.

One authority whom I interviewed identified a series of intense emotions during the initial boarding: stress, anxiety, concern, and fear. He explained that tensions ran high among civil servants and migrants and that some chaos ensued due to a lack of communication and the smugglers' efforts to thwart authorities' head count of migrants by forcing the latter to resist submission. Growing more concerned about the levels of frustration among co-workers on the response team, he asked the migrants to assume a squatting position in an attempt to calm people and place authorities in more of a position of control (Interview, Victoria, March 2001). Several photographs taken on the ships in transit to Esquimalt and at the Work Point Barracks gymnasium show migrants in this position of submission, crouched down among officers with knees bent and heads down. This image contrasts starkly with the relationship between migrants and authorities caricatured in the cartoon in Figure 3.

Once migrants were transferred and processed, the photographs show authorities searching for more clues. Agents identified carved and chalked graffiti on the boats, with one message from a practitioner of Falun Gong reading, "truth, kindness, endurance." Other photographs document horrendous sleeping and living conditions in the holds below deck, where migrants spent most of the journey in close quarters sleeping on wicker mats, eating rice cooked on gas-fueled stoves, and relieving themselves in large plastic buckets overflowing with human waste.

These photographs create images of movement around the ship through passageways, views through clouded portholes, and covered hatches left ajar. Intelligence reports compiled later based on interviews with captains and migrants would sketch people into the voids of these images.

Transfer to Land

Migrants are transferred to land at small docks in small towns where a crowd of media and other interested spectators gather to observe, and in some cases protest, the off-loading. CIC employees assist migrants off ships onto docks, and from there the migrants walk to school buses awaiting their arrival for transport to Esquimalt for processing. Despite the efforts of CIC employees to shield the migrants from the media, these images invariably come out in that day's news, to the chagrin of communications employees, who have labored to convey to the media the importance of protecting clients' identities. In one case, the media actually rigged a local surveillance camera on a dock to broadcast the off-loading live on local television.

En Route to Esquimalt

From there, the photographs document the long journey by bus to Esquimalt. Photos show migrants sleeping on buses while authorities set up secure stops along the way—a set of outhouses, food served out of the back of an RCMP van from which CIC and RCMP officers and their canine companions keep watch.

Following interception, seaworthy ships are towed and migrants are taken by ship or by bus to the Work Point gymnasium at the Esquimalt naval base. There they spend the next several days undergoing medical exams, processing, and interviews. Pictures show CIC employees handing out food, conducting interviews, and directing migrants to the various stages of processing in the gym.

Managing Bodies on the Base

Civil servants dealt first with health and safety issues; wanting to contain disease,[5] they looked more closely at migrants' bodies for clues. The following description of the process was provided by an officer who worked on site:

> We had the medical triage, number one. So they would bring them down in the buses, and they always seemed to come in the middle of the night. And we would off-load them in groups of eight just because that's how many showers we had. They would come in the back door, and we'd tag 'em. . . . We had hospital tags. They just went by a number because that was the easiest way to do it. So as soon as they walked in the back door, we'd tag 'em. They'd walk to a table with the interpreter, and we'd get their name and date of birth and all that sort of stuff, and they would go along and get photographed in their outfits that they were wearing. And then they would move to the disrobing station, where they were told they were going to have a shower. "Give us all your things, all your personal belongings." All their clothes went into big garbage bags, and they were destroyed. And then they moved into the showers, and so it was a slow process. Because the showers took ten to fifteen

minutes because they had to put . . . nit kits . . . on their hair, like for lice, and you have to leave them on like for ten minutes. And then they would come out at the other end, get a change of clothes, and then they would go sit and wait for the medical guys. It was just like an assembly line, like boat four was very efficient because we were a lot quicker, and they just moved around the gym. So in the gym, the next thing they'd go to was the medical for a full x-ray. And there were little cubicles set up. They would go into the cubicles to get examined. And then they'd come out of that and get issued a set of clothing. And then their cots were in the gym, and they'd be assigned their cots.

Then we would start the examination interviews, the next day usually. You know, it was always depending on what time we'd finished, so that the officers could sleep and all that. And then we would work really just on the interview and these forms. Some of it's intelligence gathering, the rest is just like we would do at a port of entry: What are you doing? Where are you going? Why? What do you want? That kind of stuff. We worked through that . . . and then when those were finished, or started to get finished, then the files would go to a Senior Officer. Somebody who would do all the paperwork up was called a Senior Officer. And they would sit down and do an interview with a Senior Officer. . . . That's what happened last year. We basically waited until they were all done, and then they got moved off to wherever they were going. . . . So that's what happened. There were three phases: the medical triage, examinations, and the SIO [Senior Immigration Officer] reviews. (Interview, Vancouver, August 2000)

This narrative expresses the intense focus on managing bodies as medical and migrant subjects. States perform competence through lengthy processes of identifying, categorizing, observing, testing, numbering, and recording.

Inside the gym, makeshift rooms for interviews and medical exams come into being with the hammering of plywood, the hanging of a curtain to protect privacy, and a sign announcing the place for exams. Rapid construction inside the gym and the addition of trailers as temporary offices for authorities outside the gym underscore the contingent nature of processing at the Work Point gymnasium.

Uninformed as to migrants' identities and unable to distinguish between enforcers and clients, CIC workers numbered migrants with bands around their wrists and with large numbers in black magic marker on their backs. Like the image in Figure 5 from a poster circulated by the Canadian government asking authorities and citizens to "keep watch," many photographs in this album show tattoos of birds, dragons, swords, and tigers on arms and backs.[6] Civil servants read bodies as texts and look specifically at tattoos for clues to a history of crime and triad involvement.

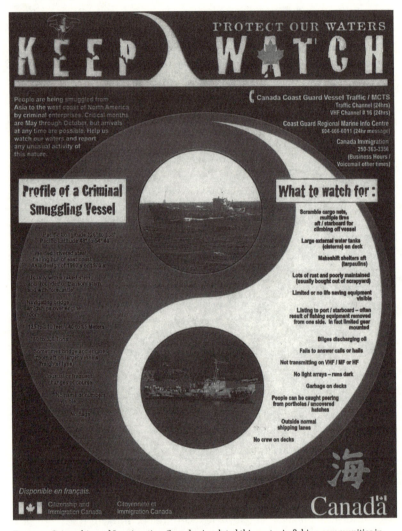

Figure 5. Citizenship and Immigration Canada circulated this poster in fishing communities in British Columbia during the summer of 2000, asking them to "Keep Watch" and "Protect Our Waters" from the arrival of boats from China. Courtesy of Citizenship and Immigration Canada.

The images portray collaboration among governmental partners. Over time, a wide range of people became involved at Work Point Barracks: legal counsel for potential claimants, provincial authorities who become guardians for the minors, corps commissionaires hired to guard the site around the clock, human rights monitors, interpreters, information technology

(IT) staff, and a small medical staff consisting of doctors and one full-time nurse.

The site expanded beyond Work Point gym over time, with trailers extending well beyond the limits of infrastructure by September, after the fourth boat had arrived. Each new boat arrival witnessed the addition of new trailers to serve as offices for legal counsel and CIC. There were also more outhouses set up, and a paved road providing direct access to the gym was added (and later removed at great expense by CIC). As capacity and optics of the response grew, so did its footprint in Esquimalt and the discontent of the largely residential neighbors.

This image of an operation in progress speaks to the contingent nature of responses to human smuggling by sea and the creative responses devised by civil servants lacking policy to guide practice. What began as a small operation in a gymnasium developed into a full-blown, if provisional, processing and detention center, with the rental of trailers to house offices, the arrangement of parking, and the hiring of corps commissionnaires to secure the perimeter around the gym.

The Disjuncture between Policy and Practice

The narratives relayed by bureaucrats related only partially to written policy, organizational structure, or prescribed roles and responsibilities. Bureaucrats' reflections on the 1999 response related more closely to the *gap* between that which had been written on paper and that which came into being through daily practice. State practices came alive in this space between formal policy and ad hoc arrangements. Policies and mandates represent the more superficial expressions of the nation-state, perhaps the most visible narratives of state activity. And it is often within the realm of policy and mandate that social scientists conduct their work (Heyman and Smart 1999, 15). Written policies, however, tell only partial stories—idealized versions of what *might* be or what *should* happen.

The frequent characterization of this work as "policy on the fly" suggests the provisional nature of the response and raises questions about the precarious nature of access. Since written policy provided little guidance as to policy or procedure regarding the interception of smugglers at sea, CIC employees themselves created policy at a regional level as they proceeded. They did have to comply with the law. They often characterized law as a barrier to be subverted rather than a guiding force in day-to-day work. As one person said, "Most of us knew that the policy says that. But you've got to get your job done" (Interview, Vancouver, August 2001). The power of the state to alter time and space to accomplish goals in spite of legal and policy directives suggests that, far from diluted, the effects of state power are concentrated along the margins of sovereign territory.

In the case of the 1999 arrivals, policy did not actually exist; it was written retrospectively in the ensuing months and years to explain what had taken place (Interview, Vancouver, August 2000). According to one official interviewed, "The experience with marine arrivals in 1999 identified concerns that our current detention policies and legislation do not provide support or guidance to deal with large-scale organized human smuggling activity in Canada."

In 1999 the only policy designed to structure a marine response to smuggling had been written years earlier in response to the arrival of 174 Sikhs on a Chilean ship, the *MV Amelie*, in 1987 off the coast of Nova Scotia. Following the arrival of the *Amelie*, the migrants made refugee claims and were held in a gymnasium at the Canadian Forces base Stadacona (Singh 1994, 140–43). Parliament was called back into emergency session to develop a response. That response became policy, but was written as a sunset clause in the Immigration Act. The clause was put in place in response to a hunger strike carried out by one senator in protest of the placement of restrictions on refugee claims. It was set to expire six months later, and it foreshadowed a broader unease on the part of the government in committing to long-term policy in this area. By the time the boats arrived from Fujian Province eleven years later, the policy provided lessons from the past, but offered no legal imperatives applicable to the present.

Nonetheless, local managers *had* prepared operational procedures, devised collaboratively at the regional level among federal partners during tabletop exercises in the spring of 1999. So in the absence of national policy, but with a rehearsed modus operandi in place, the integrity of the frontline response to the boat arrivals depended largely upon trust, personalities, subjectivity, politics, leadership, hard work, and collaboration among institutions during what quickly became a time of crisis with no defined procedures for the federal government. Amid social interactions and messy processes *within* the bureaucracy, civil servants colluded, collided, and successfully collaborated with other institutional actors such as food and equipment suppliers, interpreters, media personnel, and refugee lawyers.

Disjunctures between policy and practice did not end once first responders and migrants had, at long last, reached dry land. On the water, CIC employees had to adapt to working with members of the military, both operationally and culturally, though lacking its hierarchies.

Just as the response on the water required flexibility and the ability to deal with the unexpected, such as escapes from the beach and chases at sea, so did the ensuing months of processing and detention in BC. Once the first boat had been intercepted, CIC had twelve hours to transform Work Point Barracks into a makeshift site for processing and detention. Managers quickly devised operational procedures to bathe, clothe, and process people, while immigration officers conducted interviews and medical personnel conducted examinations. Intelligence officers worked to learn the identities, objectives, and group dynamics of the migrants. Some detainees misrepresented their identities and ages initially, probably under the guidance of the enforcers. It was challenging for officials to confirm identities because few migrants carried identity documents. It also took time for authorities to identify enforcers. CIC faced language barriers and rushed to find interpreters who understood the two dialects used by the migrants (who spoke limited Mandarin, the language of their primary-school education). Another challenge was feeding people who were malnourished and dehydrated, having been at sea with insufficient food and water for weeks.

In the absence of written policy, law played an even more fundamental role in each operational decision than it would have had relevant policy already been written (and thus already approved by legal counsel). The procedures and the physical layout of the Work Point gymnasium at Esquimalt evolved and became more efficient and legally correct over time, as CIC reduced the number of days that migrants would spend at the naval base from fourteen. In the meantime, high-ranking officials such as assistant deputy ministers and lawyers at National Headquarters guided RHQ through the response with daily conference calls that often lasted a couple of hours. Often over twenty people would be involved in these calls, with employees from various branches represented, including intelligence, enforcement, communications, and legal services. Frontline responders and the middle managers mediating communications between BC and Ottawa often expressed frustration with the slow pace at which Ottawa interpreted policy and issued commands. They also often expressed antagonism toward lawyers and policymakers alike, who they believed did not understand the realities and challenges of the frontline regional response, especially the limited temporal windows within which they needed to act. As all parties worked to develop a tactical response, the interpretation of, and debates surrounding, legal issues were often front and center in disputes between RHQ and NHQ.

At the time this appeared to be a temporary void in policy that most people interviewed believed would eventually be filled. One bureaucrat told me, "Policy loves a vacuum." Following the boat arrivals, for the purpose of guiding responses to future arrivals, policymakers in Ottawa did indeed *draft* a National Marine Policy Framework retrospectively in light of what took place. Employees from various federal departments involved on the water and at the base in British Columbia eagerly awaited the drafting, circulation, and signing of this document, because they needed the certainty and support of nationwide policy in order to prepare for future arrivals. Policy makes sense of procedures honed over time, and is more coherent in its final iteration. For this reason, policy may prove more helpful in making sense of the past than in providing guidance for the future. To understand why, it is useful to recognize the range of challenges confronted by civil servants in their efforts to quell undocumented migration with what they felt was insufficient guidance from written policy.

The Challenges to Seeing Human Smuggling Like a State

In 1999 there was a dearth of empirical information about the organization of the boats from China and a need for the performative state to fill the gaps by providing an explanation about what had happened: why the boats came, who the migrants were, and how they fit into the work of CIC. The narratives that emerged in the press foregrounded the involvement of "nefarious" individuals in transnational organized crime. I spent many hours meeting with a mid-level bureaucrat who was knowledgeable about the department's work on organized crime in BC. After mining his own institutional memory to complete this history, he subsequently told me not to believe any of the narratives circulated by his colleagues about organized crime. Although many people within the bureaucracy believed in this narrative of human smuggling, none could actually substantiate it with evidence. This same bureaucrat implicated himself and his colleagues in making easy assumptions about transnational organized crime and smuggling. In his assessment of the correlation drawn too quickly, he remarked, "If it looks like a duck, walks like a duck, and talks like a duck, well then it must be a duck." In other words, the declaration that transnational organized crime had facilitated this smuggling enterprise was an easy narrative promoted by leaders and communications employees alike. He suggested that Ottawa and communications employees had circulated a "smoking gun," an explanation of the problem where there might not yet have been any explanation known to CIC. Indeed, despite a general lack of empirical data, states contend with the demands of the public and the media for information and accountability. This employee was already in the process of interrogating those phrases that referenced transnational organized crime as they had been deployed in the public narratives of the marine arrivals, performed with confidence by his co-workers in communications.[7]

Respondents repeatedly cited as their greatest challenge the lack of empirical knowledge about the organization, recruitment, and experiences of smugglers and their clients. Most empirical evidence available is garnered through interceptions, intelligence, inland refugee claims, and, to some extent, research (e.g., Chin 1999). Because human smuggling is a phenomenon glimpsed by the state only at particular moments, information is scant. In interviews, government officials repeatedly articulated a need for more information about how people come to be smuggled or trafficked, and about what happens to them over time. As a result, nation-states are perpetually choosing reactive approaches in the form of interception and deportation. In fact, most smuggling simply eludes the government, as intended. It goes unseen, and this invisibility perhaps sets the stage for the hypervisibility of interceptions.

A primary area where knowledge is lacking relates to the geography of movements. Distance compounds the lack of knowledge about smuggling. While states know little about the geography of smuggling networks, smugglers, in contrast, appear to have intimate geographical knowledge of global staging areas and of "international weaknesses in regimes of migration control" (Salt and Stein 1997, 474). In the case of Fujianese migration to British Columbia, smugglers had knowledge of the logging roads that led to remote locations on the rugged western coast of Vancouver Island where migrants were taken, the nuances of security procedures at Vancouver International Airport, and the routes and schedules of commercial trans-Pacific container ships. Smugglers are also highly adaptive, able to shift methods and routes quickly, and willing to disregard the law and the constraints of borders. States, meanwhile, work more reactively from a distance, trying to make sense of and impose order on transnational movements by connecting the dots between clues learned in interviews.[8]

Little is known about the locations and conditions where migrants boarded the boats and traversed transit points to final destinations. Whereas Paul Smith argues, "[H]uman smugglers have minimized the importance of geography" (1997, 17), I contend that smugglers *maximized* the importance of geography. Smugglers use geography to their advantage. The success of their work and, as a result, the success of the work of law enforcement authorities rely on recognizing the role of localized geographical knowledge. Like the networks connecting cells of terrorist groups globally, smugglers pursue routes that elude governments. Their routes are always changing (Salt and Stein 1997); geography becomes the cover. Like terrorists, they are not backed by federal governments.

Following the 9/11 attacks and the discovery that the al-Qaeda terrorist network was behind them, Fouad Ajami commented one evening on the *Charlie Rose* television show, "Now there is a geography to the terror."[9] Likewise, the 1999 boat arrivals alerted Canada to a geographical underpinning of smuggling routes that centered British Columbia in the Pacific Rim. Yet these geographies tend to emerge momentarily and partially from the intelligence gathered during interception and the ensuing refugee claimant process. The boats that arrived in 1999 were not marked by any symbol of national belonging; this fact at the time rendered them visible to the federal government, but more generally allows smugglers to operate in a manner invisible to states. Smugglers work transnationally by changing the locations of bases of operations, staging areas, and subcontracted labor (Chin 1999). They change geographical routes and methods. They remain a part of the underground economy. They operate opportunistically in entrepreneurial fashion.

Various aspects of the daily work of civil servants inhibit their ability to fill these empirical silences and geographical distances. The first is bureaucracy itself. Bureaucracies are inherently large, slow, complex administrative networks that compartmentalize the division of labor in such a way as to sometimes complicate the sharing of knowledge and the taking of action. Referring to myriad factors that inhibit a faster response on the part of the government, a nautical metaphor once provided to me by a bureaucrat involved in these issues seems appropriate. He noted that if the human smugglers are akin to a slick speedboat moving craftily across international waters, the Canadian government is like a tanker trying to catch up (Interview, Ottawa, March 2001).

Furthermore, bureaucrats operate in political contexts that shift and pose constraints, depending on the political climate. There are moments when the climate calls for stronger enforcement measures and times when it does not (see Nevins 2002). When human smuggling was not in the news, political leaders were more hesitant to remind the public of CIC's enforcement mandate in the context of their desire to facilitate, increase, and promote immigration to Canada. This point was made repeatedly in interviews because it frustrated civil servants working regionally on the front lines. They were challenged by the reluctance of senior administrators to commit sustained resources to prepare a response. The state, hollowed out by neoliberal policies, grows ever weaker in its ability to prepare for emergency responses, particularly when the political will is lacking. This dearth of financial and human resources was altered dramatically following the 9/11 terrorist attacks, with the creation of the Canadian Border Services Agency (CBSA) in 2004, a separate entity from CIC with an enforcement mandate. One respondent working in the area of smuggling referred to this separation as "the divorce" (Interview, Vancouver, December 2004), a reference to the tensions surrounding such dramatic change to a federal bureaucracy.[10]

Bureaucracies also face financial constraints. Departments and other institutions struggled over who would pay for various aspects of the 1999 response, from interception to long-term detention. CIC invested significant funds immediately to create the infrastructure for processing and short-term detention at the Esquimalt naval base outside Victoria: trailers, outhouses, barbed wire, corps commissionaires, food, and medical equipment. Eventually, every expense would be justified to auditors, although the cost—estimated at $36 million for CIC—far exceeded the entire annual operating budget for the region.[11]

Another significant factor compounding the logistics of responding to smuggling was the handling of the media, a constant companion. Respondents frequently cited these interactions as the most stressful aspect of the response to the marine arrivals. The ways that states "see" high-profile human smuggling movements are intimately bound to the ways that the public learns about smuggling movements as portrayed by the media. CIC invested substantial resources in communicating this issue to the public through mainstream media outlets, and this relationship with the media is addressed in chapter 3.

If government processes seem slow in comparison to human smuggling operations, those that involve the law and lawyers were viewed among CIC employees as the slowest of all (Interview, Vancouver, April 2001). As one bureaucrat said, "As for legal services in Ottawa . . . they are always slow and reactive" (Interview, Vancouver, April 2001). CIC employees frequently expressed frustration with a variety of laws that they believed inhibited their response, including those shaping the refugee claimant process, investigative processes, and international actions on the ground in other locations such as mainland China and Hong Kong. One manager expressed his frustration with Canada's refugee determination process: "We have no control over the process. Efficiency has no connection to the system we set up. You couldn't have set up a more inefficient, complex system" (Interview, Vancouver, April 2001).

Nation-states also face constraints in the international and national laws that guide and limit their response to human smuggling. CIC employees routinely play what respondents called "quasi-diplomatic" and "quasi-enforcement" roles abroad, where they work more proactively. Immigration Foreign Service Officers facilitate migration to Canada from posts abroad; Immigration Control Officers (ICOs) suppress immigration to Canada; and Airline Liaison Officers (ALOs) police airports. In interviews, ALOs described the sensitive position that they occupied when they supported interdictions without jurisdiction to make arrests. They worked diplomatically with local authorities by passing along information and encouraging them to act on it (Interview, Hong Kong, May 2001).

CIC employees also found themselves concerned with geopolitics. In responding to marine arrivals, the Canadian government pressured China to strengthen enforcement. Canada was simultaneously careful, however, not to embarrass or otherwise taint its relationship with the People's Republic of China (PRC), a country with which it is intricately connected in geopolitical terms, linked through trade and immigration (see Smith 1997, 17). During the same year that the boats arrived, Canada sent a trade commission to China, and China had begun the process of joining the World Trade Organization. The boat arrivals and subsequent refugee determination process thus unfolded at a time when Canadian officials opted not to embarrass the PRC on its human rights abuses, which have been the basis of most successful refugee claims among Chinese migrants in Canada.

The constraints presented by international law and international relations between states mean that the response to human smuggling requires sensitive practices in diplomacy, an area in which CIC does not normally work. In 1999 and 2000, however, Minister Elinor Caplan and assistant deputy ministers traveled to China to meet with government officials. Additionally, even high-level bureaucrats working regionally in BC found themselves interacting with Chinese authorities to negotiate identity documents such as birth certificates as well as travel documents for repatriation. Some interviewees referred to this as "quasi-diplomacy," conducted by bureaucrats who had not been trained in diplomacy.

Still others drawn outside of their routine responsibilities categorized their work as "quasi-enforcement." While CIC did have an enforcement mandate in 1999, it tended to be limited to the work of those in the enforcement branch. In responding to the boat arrivals, however, a large and diverse array of CIC employees was pulled into the enforcement mandate to play new roles. Accountants and employees in the field of information technology (IT), for example, normally based in Vancouver, played a crucial role in the interception of boats *on the water*—not because of their accounting and IT skills, although they supported the response through their normal work responsibilities as well, but because they spoke fluent Mandarin. This particular challenge thus required civil servants to work well outside of their normal fields of responsibility in a bureaucracy that has had its human and financial resources reduced since the early 1990s.

Nation-states face conflicting mandates with regard to immigration, and these manifest in the day-to-day work of immigration departments. Even with the creation of the CBSA, human smuggling still does not fit neatly into one federal department or mandate. Some states overcome this challenge by separating enforcement from immigration, or by separating smuggling from immigration. The United States chose the latter course in 1993 when President Bill Clinton directed the National Security Council to combat smuggling, which later led to a coalition of federal agencies working on the issue (Smith 1997, 2).

The response to human smuggling also requires collaboration among various governmental and nongovernmental institutions to enact a successful response at all stages, including interception, processing, detention, facilitation of refugee claims, resettlement, and repatriation. CIC collaborated with a broad range of institutions in the response. Some collaborations drew on long-standing relationships between CIC and refugee settlement agencies, while others called upon CIC employees to build relationships with institutions such as the Red Cross, provincial players such as the Solicitor General who oversaw detention, and federal departments such as the Department of National Defence, to name a few. All of these relationships brought together institutions with distinct mandates, cultures, policies, and frameworks.

One of the areas where this collaboration became particularly sensitive was intelligence. Intelligence gathering, while crucial to the government response to smuggling and among the few ways for states to become more proactive, also poses cross-institutional challenges. In 1999 intelligence officers worked from the moment of interception to fill in details in the history of the arrivals. They asked questions pertaining to how people had boarded the vessels, who had assisted them, and whether and where they had stayed in safe houses. They analyzed data gathered from immigration interviews, letters, and phone calls to and from the detained migrants over time. Founded in 1965 to assess the threat of organized crime to Canada, the intelligence division worked primarily abroad. In

1999 its personnel were still distributed throughout the department both regionally and functionally, with intelligence officers working on topics in somewhat haphazard fashion according to their geographical location. A 1996 review of the department's intelligence function found: "There is no functioning, effective departmental process. Instead there is a series of Branch processes which sometimes collide and, at best, fail to complement one another." Intelligence work was further inhibited by the slimming of the department in the early 1990s, which coincided roughly with its being stripped of its investigative status. Additionally, immigration intelligence has not always been embraced or integrated into the broader Canadian intelligence community. Such "intelligence silos" made it difficult to share information among federal departments within Canada and across international borders.

"Keep Watch"

With empirical knowledge lacking and the state distanced from localized knowledge, institutional collaborations sometimes take surprising forms. In the spring following the boat arrivals, RHQ circulated a poster (shown in Figure 5) to fishing communities along the western coast of Vancouver Island and on the Queen Charlotte Islands.

Through its text, the poster conveys a message to Canadians to "Keep Watch" in order to "Protect Our Waters." It explains that "criminal enterprises" are smuggling migrants to the western coast of BC and invites people to contact the Coast Guard or CIC to report the observation of "any unusual activity of this nature." The text goes on to describe the "Profile of a Criminal Smuggling Vessel" with a series of physical characteristics such as the expected size, condition, coordinates, markings, and navigational behavior patterns of migrant ships. A second column lists "What to watch for," a listing of additional details about the vessels and the people they carry, noting that there will likely be no crew visible on deck, but that people "can be caught peering from portholes/uncovered hatches." Photographs of two of the ships intercepted in 1999 accompany this text, along with a series of orientalist cultural symbols.

Like the text and photographic images, the symbols draw on a Canadian national lexicon and reveal something about how this state sees smuggling. A red maple leaf covers the "A" in WATCH and an osprey perches on the "E" in KEEP, ready to swoop. An enlarged version of the poster also shows the Canadian maple leaf in the eye of the osprey. The symbol of a wave moves between the words "Keep" and "Watch" and accompanies the

Chinese symbol for "ocean." Like written policy, this poster in practice constituted more of a record than a forecast. No additional cargo ships transporting migrants in this manner were intercepted. Smugglers again proved themselves more flexible than the state.

This poster distills down to a set of images the ways in which the state set its sights on human smugglers and the importance of visuality during what bureaucrats referred to as "boating season."[12] Its intent supports the argument that the boundaries surrounding state practices are fluid, inasmuch as CIC relies on collaboration with a host of other parties and institutions in the response to smuggling. The poster relays a daunting reality for a government without much capacity for marine response: the need to rely on others, and especially on local knowledge, in order to detect smugglers. The larger characters in the text are typeset in the same broken black and white text used for the logo for the popular American television show *America's Most Wanted*, which invites a national audience to assist in solving crimes that remain open, the whereabouts of the fugitives unknown. But here there are no snapshots or sketches of smugglers, only the shell of crimes committed. Local knowledge is needed where the state cannot see, and to attain that knowledge, bureaucrats need to build relationships and trust with various groups and institutions. This reliance on others includes other federal departments in Ottawa and spotters on the ground in British Columbia and abroad.[13]

The poster also affirms Amoore's emphasis on the role of visuality in securing sovereignty, a role that is increased during highly visible marine interceptions evoking fears of invasion. The intense focus on ships signaled exceptionalism by diverting attention from the normalcy of unauthorized entry, routine and largely invisible at other ports of entry such as airports and land crossings. This focus also intensified associations with criminality and surreptitious, illicit entry, narratives that would ultimately undermine the integrity of access to the refugee determination process.

Detention and Deterrence; Visuality and Access

States have devised a variety of strategies to combat human smuggling. They vary in terms of cost efficiency, desired effects, and unintended consequences. CIC pursued a range of strategies in 1999, but invested most of its resources in the detention and removal of rejected refugee claimants.

Detention entails a variety of substrategies. Because smugglers receive payments over time once clients reach their destination and amass earnings in the years to come, detention or "detention and removal" functions as a deterrent intended to disrupt smugglers' revenues by preventing clients from reaching their destination. Intelligence and enforcement employees in immigration also use the time during which migrants are detained to learn about the journey and smuggling operation. In 1999 they monitored telephone calls, mail, and the individuals who posted bail in order to find out the people and places to which the detainees were connected.

There are several problems with detention. In addition to the cost, the logistics of detaining on a large scale dominated the workload of people in CIC. As with other assets required for marine interceptions (such as boats), CIC had neither the infrastructural capacity nor the funds to support a program of large-scale, long-term detention in BC. In the years following the 1999 arrivals, RHQ entered into a variety of contracts involving the renovation and reservation of space in provincial prisons, such as the Vancouver Island Regional Correctional Centre in Victoria, to detain migrants. They had to withdraw from these contracts when maintaining them proved too expensive. The lack of commitment to longer-term contracts reflected the indecision within the department about whether or not to build this capacity on a national scale. It also illustrates that detention is a tricky, expensive, and long-term commitment—one not entered into lightly.

Another problem with detention is the peril of incarcerating migrants who are refugee claimants for relatively long periods.[14] In addition to suffering a difficult journey, many had experienced difficult life circumstances in Fujian. The additional stress of incarceration and uncertainty further threatened their mental health and wellbeing, and prolonged the trauma associated with their harrowing journey across the Pacific.[15] Resistance in the form of riots, hunger strikes, and suicide attempts took place among the claimants housed in BC provincial prisons, as has occurred elsewhere when asylum or refugee claimants have been detained for long periods of time (Steel et al. 2006).

Another strategy involves more diplomatic measures. Canada was proudly vocal about its leadership role in drafting and ratifying the UN Protocols (United Nations 2000) to combat trafficking (Interview, Ottawa, March 2001). CIC employees anticipated, however, that the Protocols would have little effect in terms of governance in receiving countries such as Canada. The Protocols' primary objective was to apply international diplomatic pressure on source and transit countries where smuggling and trafficking movements originated, to increase policing and penalties.

Canada also pursued more informal diplomatic dialogues where CIC entered the role as "quasi-diplomat," conducting work generally undertaken by the Department of Foreign Affairs and International Trade (DFAIT). Most of this quasi-diplomatic work entailed senior-level managers and the minister, deputy minister, and assistant deputy ministers traveling to China in the months following the boat arrivals. This was part of a publicity campaign to meet with government officials and to discourage potential clients of smugglers directly. In a bimonthly newsletter distributed by CIC called *Vis-à-vis for the B.C./Yukon Region*, a short piece on the minister's upcoming visit to China in early 2000 summarized the message that she would send on her trip: "Snakeheads are lying to you, they are cheating you, you put your lives at risk when you get on those boats or into a container, we have found people dead" (*Vis-à-vis* 2000, 1). CIC also posted such messages on billboards in the regions in Fujian that migrants had left to travel to North America.

People lower in the hierarchy of the bureaucracy, in enforcement and intelligence, also played important quasi-diplomatic roles in their travels abroad to communicate and collaborate with colleagues at their level in other governments. This is the type of work that ALOs and ICOs abroad do on a daily basis, but mid-level managers and senior people in intelligence and enforcement also traveled to Australia and the United States on a number of occasions to meet with their counterparts to learn and share strategies.[16]

Another strategy entails prosecution and stiffer punitive measures— incarceration and fines—such as those included in the Immigration and Refugee Protection Act. There are two main critiques of prosecution as a strategy to curb smuggling. First, enforcement officers express frustration with their binding relationship to a justice system that they perceive as too slow to contend with the speed and dynamism of smugglers. In order to prosecute enforcers, investigators and lawyers must meticulously investigate and assemble a case against them by working within the frameworks of multiple federal departments (the Department of Justice, CIC, and the RCMP). This is costly in financial terms, but the main investment that frustrates those involved from an enforcement perspective is the amount of time required to build a case. The second critique is that despite the threat of stronger punitive measures that accompany successful prosecutions, they are simply not a sufficient deterrent compared with the lucrative profits to be gained from the business of human smuggling. Still, Canada did pursue this strategy by prosecuting the Korean members of the crew on the second boat[17] and by specifying harsher punitive measures for smugglers in the new Act.

A range of more proactive deterrent strategies is available to governments, and CIC has also pursued some of these. A slightly more effective, if more ad hoc and controversial, strategy has been the tightening of front-end controls. Like other nation-states where smugglers operate, Canada increased its capacity for interdiction abroad by posting additional Immigration Control Officers in source countries such as China (Citizenship and Immigration Canada 2001). They also placed ALOs in transit countries. These authorities are part of an Immigration Control Officer Network "aimed at better protecting the integrity of the refugee determination process and the immigration program as a whole" (Citizenship and Immigration Canada 2001).

Interdiction and Other Transnational Sovereignties

States trade transnational enforcement strategies with the understanding that smugglers also study strategies and prioritize according to cost efficiency and desired outcomes. Canada's transnational enforcement effort extended far beyond the continent. During the year following the boat arrivals, CIC sent teams from BC to study detention practices in Australia, prosecution in Europe, and interception in the United States. These foreign immigration departments also invited Canadian civil servants to visit in order to instruct them in "best practices" in Canada.

While immigration policies have often been thought of as domestic, the work of ALOs and ICOs abroad illustrates that states, like migrants, behave as transnational actors on the landscape of human migration. In their daily work, they are pushing Canadian borders farther afield. Through actions abroad, including interdiction, intelligence gathering, and diplomacy, the state operates as a transnational actor; traditional boundaries to sovereign territory dissolve as enforcement practices extend abroad. Through collaboration and the sharing of information with other governments and private corporations, states reach across borders and oceans to manage migration long before people arrive at its ports of entry.

The Canadian state thus behaves more transnationally in its enforcement of immigration policies, particularly with tighter "front-end" restrictions. These trends involve a range of strategies, including increased security checks for those who apply for refugee status once in Canada and increased interception abroad of would-be refugee claimants. Between the establishment of the ICO network in Canada in 1989 as "a control and enforcement liaison officers network" and a review conducted in 2001, the number of positions has grown from ten to thirty-four (Citizenship and Immigration Canada 2001).

In Hong Kong in 2001 I interviewed ALOs from Canada, the United States, Australia, the United Kingdom, and New Zealand, who monitored and intercepted smuggling movements and potential refugee claimants. Though working beyond their jurisdiction, they were able to gather intelligence, identify trends, and share information with local authorities. In addition, they fostered informal working relationships with a network of international colleagues and with airlines, which enabled them to increase vigilance among airline staff, check documents, and pull passengers from lines and flights to conduct interviews (Interview, Hong Kong, May 2001). They sometimes roamed the airport themselves to screen those boarding particular flights, and invested as much time training airline staff to look for false documents and to call when they encountered a suspect individual (Interview, Hong Kong, May 2001). A review of the network found that it had "improved relationships with other countries facing similar immigration issues," "reduced the number of inappropriately documented passengers," and "decreased the number of illegal entries to Canada" (Citizenship and Immigration Canada 2001)

After the boat arrivals, CIC also placed an additional ICO on the ground in China. With the Smart Border Accord, Canada increased its total number of immigration officers working overseas to more than eighty-six; the United States deployed an additional eighty-five "new temporary officials," with forty new officials deployed permanently (United States Department of Transportation 2005, 4). ICOs and ALOs reduced the numbers of migrants traveling to North America with false documentation and the intention to make an asylum claim. Refugee advocates argued in response that this increased interception abroad pushed those seeking asylum to employ the services of human smugglers (United States Committee for Refugees 2000b).

Another, more transnational strategy pursued in Hong Kong in the years following the boat arrivals was the decision to freeze the financial assets of those under investigation for involvement in smuggling, and then track the connections to the money (*New York Times* 2003). Each of these strategies relies on more extensive gathering and analysis of information. The controversial practice of interdiction enables states to overcome some of the distancing of bureaucracy. In some ways, civil servants are able to behave more like smugglers: to operate beneath the radar, to become more adept at seeing the local, to acquire knowledge that translates quickly into power.

The years following the 1999 boat arrivals saw substantial changes in Canada's domestic policies and participation in regional agreements in

North America (Gilbert 2005; Nichol 2005), leading to dramatic revisions in its border enforcement strategies and shifted patterns of undocumented migration. In a post-9/11 security environment, ideas rooted in the discussion of free trade arrangements and globalization in the 1980s and 1990s have come to hold more currency in public discourse, policymaking circles, and multilateral dialogues. The Canadian government pursued a strategy of harmonization based in part on the premise of perimeter theory (Koslowski 2005). This move reconceptualized the borders around North America as a whole and brought governments into more cooperative planning around immigration and asylum policies, not unlike the processes surrounding immigration and border harmonization in the European Union.

Deputy Prime Minister John Manley and Governor Tom Ridge signed the Smart Border Declaration and 30-point Action Plan in 2001. This wide-ranging action plan proposed concerted efforts to integrate migration to and through North America with the expansion and coordination of biometric technologies, permanent residency cards, inspection systems, visa policies, maritime security at terminals and seaports, immigration and passenger databases, and other measures. Of particular importance in relation to the impact of these efforts on human smuggling is the implementation of the Safe Third Country Agreement (STCA) between Canada and the United States in December 2004. A form of harmonization of asylum practices, safe-third-country agreements restrict claimants who have transited through a "safe third country" between sites of origin and destination from making a claim in the final arrival location; in other words, the claim must be submitted in the first country with a claimant system that the migrant enters. The implementation of the STCA in North America significantly reduced the number of claims made in Canada, because so many claimants had passed through the United States en route to Canada. Those people who manage to enter Canada can make claims inland, but are prevented by the STCA from doing so at a port of entry along the land border. Following its implementation, those making claims at land border crossings were turned back to the United States.

CIC employees working in the field of human smuggling were frustrated. Some mentioned that Canadian Border Services Agency (CBSA) officers had not necessarily been trained in the sensitive issues involved in working with refugee claimants. They believed that the disparate mandates of CIC should have remained integrated. CIC employees themselves pointed out the geographic hypocrisies of the STCA: that it affects those who arrive by land but not by boat or plane. While some officials higher up in the administration denied this, one civil servant interviewed in BC insisted that this agreement would lead to a boom in the human smuggling industries operating in and around Canada (Interview, Vancouver, December 2004). Little information exists to explain why the numbers of claimants dropped dramatically at first, in 2005, and then rose again, but interviews with employees of NGOs on both sides of the border suggest that human smuggling has increased.

The new CBSA combined border functions once shared between Canada Customs and Revenue Agency (its Customs program), Citizenship and Immigration Canada (its Intelligence, Interdiction and Enforcement, and Port of Entry immigration programs), and the Canadian Food Inspection Agency (its Import Inspection at Ports of Entry program). Within the field of immigration where smuggling falls, it separated the mandates of selection and settlement (still mandates of CIC) from border enforcement (now the work of the CBSA). Like refugee advocates, some civil servants confronting human smuggling viewed these as short-sighted strategies. They cited the risks of harmonization to the protection of Canadian values and interests surrounding immigration and refugee programs as well as its objectives in border enforcement (Interview, Vancouver, December 2004).

Additionally, like many other immigrant-receiving states, Canada passed anti-terrorism legislation following 9/11, Bill C-36, which further enabled the transnationalization of policy through information and intelligence sharing, the integration of databases, and other cross-border policies.

Seeing and Being Seen as Sovereign Sensibility

This chapter has conveyed, in James Scott's terms, how nation-states see undocumented migration in the form of human smuggling and work the interstitial spaces between policy and legality in which they operate. It drew connections between (1) asylum and protective mechanisms and (2) smuggling and enforcement measures, looking at the clash of these mandates as a source of crisis for states. The chapter has also engaged Louise Amoore's notion that visuality functions as a key sense and affective register to enact sovereign power to "secure the nation." Through such affective registers, state and media engage the public, and visuality becomes central to the advancement of enforcement agendas along borders during times of crisis. If Scott's state attempts to manage from a distance, Amoore's state evokes power through distant and distancing visual representations.

The geographies of the nation-state encountered in this chapter involve the dissolution of boundaries, the devolution of work to non-state actors through private and public partnerships, and the necessity for collaboration.

Viewed through interventions in transnational migrations, states work as diverse networks of employees that interact with a variety of institutions to enact federal immigration policies and mandates through practice. The anxious encounters between states and migrants conjure international borders wherever they occur in ports of entry—what Mark Salter calls "delocalizing the border" (2004, 10). They demarcate the increasingly dispersed margins of the nation-state where power differentials are reproduced amid moments of crisis in the encounters between authorities and migrants.

States have undergone a shift in their confrontation of human smuggling. They are now acting more transnationally. This chapter conveyed how nation-states see, using the metaphor of sight to explore the geographical relationships between states and smugglers. Both parties look upon the global landscape from varying proximities. As exemplified by the "Keep Watch" poster, states see the success of future interceptions as bound up with the vigilance of others interested in policing borders, which requires relationship-building, communication, and a sense that individual citizens can participate in border policing as they go about their daily lives. The view of the state is formed through the daily work of civil servants who share information through a variety of informal networks. The end result resembles a patchwork quilt of images, rather than a comprehensive portrait of what is often called "the environment scan" for smuggling. Through the everyday practices of civil servants, human smuggling looks different around the globe. In Hong Kong it looks like the wrong accent, the wrong clothing, the wrong story, or the wrong papers in the line to board the plane. In the office tower in Ottawa for the intelligence analyst, smuggling looks like a geographical trend coming across the listserv or database. To the fisherman off the Queen Charlottes, it looks like an image in the news or on a poster.

The problem is that a smuggling movement "looks like a duck" only once. Then it transforms into something else, takes on other shapes and performs other optical illusions, such as young Chinese men walking through airports dressed in the baggy pants and other urban gear of Asian–Canadian youth (Interview, Vancouver, May 2000). In the years following the 1999 marine arrivals, while boats carrying migrants in large numbers abated, other, quieter forms of smuggling continued. There were quick shifts from boats to container ships, to air cargo holds, to boarding pass swaps in airports. During the months following the boat arrivals on Vancouver Island, several incidents involving Fujianese migrants as stowaways in containers on commercial container ships took place along the West Coast—in Long

Beach, California; Seattle, Washington; and Vancouver, BC. Two container ships were intercepted in Vancouver's port in January 2000, four months after the last boat arrival. These ships were on their way to the United States.

Meanwhile, the bureaucracy spent months—and, in the case of the 1999 boat arrivals, years—designing policies and posters to respond to what looked like a duck. Meanwhile, migrants crossed borders by other methods, such as travel by air in cargo holds and even in the landing gear of airplanes (e.g., *New York Times* 2001). In the efforts to categorize migration and to design and implement policies, the bureaucracy struggles to keep pace. Each "boating season" renews heated debates about the reasons why no more boats have come. Most senior-level officials whom I interviewed in Ottawa were quick to attribute the change to the successes of their response to the boat arrivals. But were they successful? Or did smugglers simply shift routes and methods? These were the questions asked by civil servants working on the ground in BC and China.

As states invest more resources in seeing smuggling beyond the purview of sovereign territory, sovereignty grows more transnational in daily practice. Information crosses borders through shared databases and networks of collaborating colleagues. This increased investment in resources, however, does not necessarily mean a longer-term commitment to policy and planning for marine arrivals. Often these practices are not written into policy, but rather are rendered visible in the contingent arrangements of daily practice. Ethnographic research offers the opportunity to map the daily practices of the transnational state in its efforts to identify and suppress human smuggling. In the next chapter, an ethnography of the state uncovers the contingent nature of the response to human smuggling unfolding inside the bureaucracy.

Chapter 3   **Ethnography of the State**

I BEGAN RESEARCH IN 2000, during the summer following the interceptions, when everyone in the office anticipated the arrival of more boats.
The months following the marine arrivals were a tumultuous time; they
offered an opportunity to explore the operation of paradoxical narratives
of the state as powerful and vulnerable during times of crisis. This chapter
explores the daily practices of the performative state, and ensuing chapters
will show how the exceptional practices in times of crisis outlined here give
rise to "exceptional zones" (Agamben 1998) along the shifting margins of
sovereign territory, where exclusion transpires.

On and off for several months, I participated in the daily exchanges of
office life at Regional Headquarters (RHQ), without actually doing any
immigration-related work. I sat at a desk that served as my base from which
I interacted with and observed employees in immigration. I reviewed documents and conducted interviews with employees about their roles in
response to the boat arrivals. I attempted to interview everyone involved in
the response to human smuggling, including frontline officers who boarded
the boats, officials much higher up in the administration of the department at National Headquarters (NHQ) in Ottawa, and civil servants posted
abroad in Hong Kong. I asked questions about the marine arrivals, but was
equally interested in analyzing the day-to-day operation of the bureaucracy through participant observation.

The RHQ offices were so quiet that one employee called them "the
morgue." Yet the narratives I collected contradicted popular stereotypes
of the mundane nature of bureaucratic work. Interviews were infused
with emotions surrounding the controversial and contingent response to
human smuggling. As in Steve Herbert's (2000) ethnographic encounters
on patrol with officers of the Los Angeles Police Department, the affective
domain of interactions during participant observation here underscored

the degree to which human agency drives the uneven implementation of public policy.

Immigration bureaucracies are ideal locations to analyze the creation and defining of nation-states. As Basch, Glick Schiller, and Szanton Blanc note, "Nation building processes are interwoven into the very matrix of the nation-state, embedded in its institutions, manifested in its policies and practices, and organized through state bureaucracies" (1994, 37). The work of immigration officers entails daily exercises in nation-building as they implement ways to determine national identity by deciding who belongs within or outside of the nation-state. Local immigration officers exercise administrative discretion in their work where they not only influence, but are also influenced by, regional social and economic contexts (Heyman 1995).

From the perspective of those outside of government, whether immigrants, refugee claimants, immigration service providers, lawyers, or academics, immigration departments appear to be disembodied institutions. Prospective immigrants face common experiences with immigration departments around the globe, such as long lines, opaque policies, and inaccessible decision-making processes. Decisions regarding a client's status are often made in remote administrative centers and handed down without reference to the individual decision-maker. The less accessible the decision-makers are to those whose lives they influence, the larger looms the power of the state to act without accountability.[1] As Michael Taussig (1997, 3) suggests, power expands through absence: "Could it be that with disembodiment, presence expands?" The remote nature of bureaucratic work in offices removed from sites where migrants enter sovereign territory extends state power through disembodiment.

Lawyers, service providers, and advocates who worked with the migrants from outside the parameters of government reinforced this impression when they explained the difficulty of establishing long-term working relationships with CIC. They frequently mentioned inaccessibility, secrecy, and a high turnover rate as barriers to communication and relationship building. Indeed, CIC was known jokingly among immigration consultants as "the fortress." With limited communication and collaboration between immigrants and refugee claimants, their spokespeople, and those who decide cases, an advocacy industry arises and operates in adversarial fashion with CIC. In my research, I explored from the inside how and why CIC appeared disembodied, given its central role in responding to human smuggling.

Yet, as in any other workplace, the bureaucracy is embodied. Bodies labor there and experience many of the pressures and much of the politics present in other places of work, intensified during crises. This chapter offers an ethnography of the state in its daily attempts to manage transnational migration, and the power circulating through this "management" by examining the response to human smuggling at RHQ in Vancouver. Analysis of daily state practices in the response to human smuggling complicates the monolithic characterization of "the state" and proves its institutional networks to be socially embedded. States are shaped by the local and regional communities in which they operate, which in turn shape them. Beyond this local context, states' actions reverberate transnationally.

The exclusion of immigration officers in conceptual models of migration normalizes the taken-for-granted right of the state to define categories of human displacement and to determine legality and illegality (Heyman and Smart 1999; Nevins 2002). This silence also erases differences within the state. Scholarship on immigration requires deeper contemplation of "the state," and specifically the bureaucracy, in order to demystify and deconstruct notions of homogeneity among decisionmakers. The bureaucracy is only one part of "the state," but one with all of its parts in motion as civil servants put the law, financial resources, and various institutions to work.

Most studies of immigration bureaucracies have examined what was until recently called the Immigration and Naturalization Service—or INS—of the United States and usually involved the relationship between the INS and Mexican migrants crossing the United States–Mexico border (Calavita 1992; Heyman 1995; Nevins 2002).[2] Josiah Heyman (1995), for example, conducted research with INS officers regarding their frontline work. He examined their social histories and subsequent socialization within the bureaucracy, and looked particularly at the manifestation of how these factors influenced their decisionmaking processes with regard to immigrants. Joe Nevins (2002) documented Operation Gatekeeper, a federal enforcement objective requiring the investment of several million dollars in human and technological resources along the United States–Mexico border in the 1990s. He examined the relationship between the United States and its immigrants, and in particular tracked the construction of the concept of "illegal immigrant" over time. Nevins was especially interested in the idea that border enforcement functions as a performance of the state, as national territorial symbolism where the border signals the constitutive inside and outside to belonging.

The parameters of territory and bureaucracy are intricately bound. As Heyman noted, "Bureaucratic work is internally conflictive but appears, in the single-stranded relationship to the exterior, to be definitive . . . and rational" (1995, 264). The policies of the state are enacted amid tension and difference, but higher-level bureaucrats and communications employees construct coherent narratives for the public, which tend to provide relatively little insight into what actually took place. Communications strategists thus reinforce concealment in their work to craft a unified, coherent narrative for public consumption.[3]

This chapter sustains Scott's notion of seeing via analysis of migration management that entails centralized administrative practices of classification but adds the elements of performance and performativity. Driving this chapter conceptually is the idea of the performative state. Judith Butler (1990) developed the notion of performativity as the iterational practices through which categories become reified. This concept of performativity leads to an abstract view of "the state" as an entity that takes on meaning through its perpetual reproduction in the geographical imagination through the daily work of civil servants, citizens, and noncitizens. This performative reification of the state shapes the self-conscious performances of bureaucrats regarding their work. Bureaucrats perform the idea of the state as they enact policy for audiences both imagined and real.

Civil servants act on their personal conception of the nation-state and expectations of its role in the global community, imagined through iterative processes. The state becomes a series of performances and practices that involve negotiations and power plays, made especially apparent in the response to events of human smuggling, conjured as crises at the coastal edges of the nation-state. Joane Nagel (2003, 53) identifies tension "between the performed—concrete, obvious, purposive, deliberate ways . . . and the performative—abstract, hidden, unthinking, habitual ways" in which categories are *constituted*. Entry into their daily work spaces shows civil servants' recursive processes reinforcing the performative constitution of "the state."

The Office

CIC Regional Headquarters in Vancouver sits quietly embedded in the buzz of a bustling downtown. Whereas government buildings dominate the built landscape in Ottawa, CIC's three offices in Vancouver are dispersed amid the flow of tourists and other shoppers, demarcated only by the Canadian flags outside (Figure 6).

Figure 6. The main entrance to the regional headquarters of Citizenship and Immigration Canada, British Columbia–Yukon Region, located in downtown Vancouver. (Photograph by author.)

At each office entrance a small plaque indicates "Citizenship and Immigration Canada" rather than specifying the administrative branch housed there, such as "Admissions" or "Enforcement." Inside is a space with offices around the perimeter. Some workers enjoy window views, while others occupy cubicles in the center.

Just as quiet exteriors conceal the busy operations within federal offices, other aspects of bureaucracy function to conceal identities. The understated nature reflects the aesthetic homogeneity of federal offices. Wall-to-wall carpeting connects gray walls that enclose generic cubicles. A youthful version of Queen Elizabeth looks down from the front of every office, a subtle reminder that Canada remains a part of the Commonwealth and is still led—if symbolically—by the monarchy.

Civil servants are trained to move from one department to another and function with minimal adjustment, regardless of geography. The workers I observed did little to personalize their office space. Many put up a few tasteful family photos, and there was the occasional photograph or objet d'art memorializing an unusual place or event in the field. Otherwise, the anonymity of the place extended even to the way in which people answered telephones. The usual telephone greeting was "Citizenship and Immigration Canada," unaccompanied by a branch designation that would help to distinguish the individual among his or her four hundred colleagues in the region.[4]

The microgeography of the office reflected the power structures. The Director General (DG) occupied a large corner office in a prime location with administrative assistants lined up in cubicles outside. Directly across from the DG were the offices of communications employees. This centrality reflected their role in the day-to-day functioning of the office. Various managers occupied a number of private offices around the inside ring of the floor, and the remaining employees occupied cubicles, with the exception of those working in intelligence. Intelligence analysts were the only employees with private offices, partitioned and protected by reinforced walls. This arrangement signified an attempt to protect and also to partition off intelligence workers.

The floor where I was placed opened onto a balcony above a busy street where smokers congregated for breaks. At the center of the office was a small lunchroom with a coffee pot, microwave, and water jug where some met during the lunch hour.

The office was a quiet one. Most days this silence drew my attention to the constant hum of computers. One woman recounted being asked to be quiet when she began laughing in her cubicle. She mentioned other incidents when co-workers turned down the volume of one another's phones (Interview, Vancouver, August 2000). The controlled air and tinted windows further subdued the atmosphere by blanketing the vibrant street life surrounding the building in a shade of gray, even on sunny days—which are rare enough in Vancouver.

While the architecture of cubicles does not allow for much privacy, it did foster interaction, including ritual gatherings throughout the day. Employees' daily routines started early—by 7:30 A.M.—to facilitate communications across the three-hour difference in time zone to NHQ in Ottawa—and included ritual coffee breaks around 9:00 or 10:00 A.M. and lunch around noon. Most people traveled together to nearby coffee shops for breaks. On Friday mornings, employees gathered in the conference room for coffee and donuts. The day usually ended anywhere between 3:00 and 6:00 P.M.

I sat down the hall from the DG and communications employees, alongside employees working in Information Technology (IT). IT serviced not only those regional offices close by, but also smaller, more remote offices located in the interior of the province. Calls came in from islands, airports, and border crossings reporting that a network or computer was down, and IT members, all of whom carried beepers and cell phones, would mobilize in response. They spent long hours at RHQ and all over the province. I witnessed their work in crisis-mode, along with more mundane moments, such as during one of the frequent printer jams.

IT workers contributed to the atmosphere of constant comings and goings to and from the office as people gathered for training workshops, meetings, and press conferences. Faces changed continuously not only because of these cyclical visits, but also because of a high turnover rate. It was routine for someone new to be shown around the office and introduced. People were transient because of internal transfers to other CIC offices, shifts to other departments, or assignments abroad. Employees also traveled frequently between Vancouver and Ottawa.

The geography of the workplace represents a network in action; employees circulate continuously among offices located throughout Canada and the rest of the world, linked technologically through various lines of communication maintained by the hard-working IT employees.[5] The high turnover rate and the assumption that workers can be transferred easily between offices inhibit the accumulation of geographically specific knowledge, institutional memory, and informal local networks, which are essential to a successful response to human smuggling.

Behind the bland façade of sameness of what Akhil Gupta (1995, 392) calls the "translocalities" of the state lay a fascinating set of power struggles. CIC is a relatively small federal bureaucracy with a surprisingly large role to play on the national political stage. The scaling back of what was once the much larger Department of Citizenship and Employment Canada in the 1990s left what some referred to as the "bare bones" of the department: the concentration of administrative, financial, and communications functions.[6] Fewer staff, of course, meant more work. In the words of a DG for the BC–Yukon region, one of five regional heads in Canada:

I have about 400 staff, an annual operating budget of just over 20 million dollars.[7]. . . In this region, we deliver all of the services for Citizenship and Immigration. So that includes for Citizenship, all of the ceremonies, the testing and ceremonies for all new citizens in British Columbia. Last year we did about 35,000 new citizens. It includes all of the services at the ports of entry for people that are coming into Canada as visitors: students, business visitors, NAFTA, or just tourists. It includes all of the services for immigrants that are already in Canada that are seeking to remain. . . . If they're seeking employment authorizations from the system, getting clearance, facilitating individuals who are already inside Canada . . . but wish to extend their visit. (Interview, Vancouver, August 2001)

In short, the offices were an intense if quiet space. They housed a bureaucracy that saw itself as underresourced and overworked. Maintaining the image of order and control for the public was a challenge.

The Performative State

With all of this work to do, one civil servant described his job as being a member of an office staff running madly to maintain the façade of a wall before it collapsed, an image he likened to the Wizard of Oz hiding behind smoke and mirrors. The researcher able to step behind the public façade to study the daily work of bureaucrats may stand at odds with the objective of communicating a coherent narrative. But, as Herbert argues, participant observation offers the opportunity to observe not only what people *say*, but also what they *do* (2000). This characteristic of ethnography proves essential to understanding states. Multiple agendas are at play beneath the surface, yet institutions must maintain an outward image of coherence, a clear narrative and rationale for state action communicated to the public through the mainstream media.

Social scientists have extended Erving Goffman's (1959) front-stage/backstage "presentation of self in everyday life" and fieldwork as performance (Katz 1992; Pratt 2000). Goffman suggested that as people interact, they act out roles with a desire to make particular impressions that may or may not be perceived as intended. More recently, poststructural and postmodern theorists have conceptualized identity as fluid and incomplete, forever shifting and constituted contextually. Judith Butler's (1993) notion of performativity relies on the idea of citational iterations. This concept differs from Goffman's view of the intentionality of human agency, and focuses instead on the ways identity is assigned meaning through discourse. Both views are relevant to bureaucracies where there is a hyperconscious performance of the enactment of immigration policy for the public. These conscious performances for the media translate into broader if more mundane iterative processes that are constitutive of the idea of "the state" and its power to regulate human mobility. The public's perceptions of the state shape these daily performances.

Civil servants performed their state roles differently. Often their imaginations of the state and their particular perspective on human smuggling defined their position. Operational employees, for example, expressed frustration with the amount of power wielded by the communications branch, which is charged with communicating policy to the public. The influence of communications employees extended to every aspect of life at CIC. I had initially thought that communications employees might be most willing to work with me because their job involved the communication of departmental decisions. I found the opposite to be true. They reacted to me warily, as an individual stepping behind the image put forth for the public. Intelligence officers, on the other hand, who I had imagined would be guarded with more information to protect, turned out to be more open, perhaps because their own work corresponds most closely with the research process.

The juxtaposition of employee standpoints suggests the importance of public image to the department, and to the conflicting perceptions of the need to protect or share information. Ethnographic study revealed daily practices concerned with maintaining a public image of the state being in control and an inclination to pass blame when something went awry. Respondents mentioned frequently that there was no excuse for mistakes in the public eye. Mistakes were equated with crisis and accompanied by the assignment of blame. Crises engendered by human smuggling in which the government appears inept (Greenberg and Hier 2001) disrupt the designing of a coherent image for the public, setting into motion a frenzied spiral of communications and crisis management. The front-stage barrage of unprecedented coverage in local, national, and international news outlets coincided with back-stage policy on the fly, and set the stage for the exceptionalism that ensued in claimants' access to the system amid the performative enforcement response.

The Crisis

In contrast with the quiet hum of daily life at RHQ are the extreme terms employees used to describe their work experiences when responding to the boat arrivals. These descriptions involved animated facial expressions and hand gestures which conveyed an enormous degree of emotion and the feeling of being pushed to the edge. Respondents often likened the response to times of war and natural disaster. These metaphorical descriptions of the environment involved violence, danger, and threats to the personal safety of both migrants and the authorities who put themselves at risk in out-of-the-ordinary, out-of-the-office work experiences.

In the following interview excerpts, the DG compares his life during times of crisis and "normal" times, a distinction that he and others made often.

> On an average day here . . . the first thing I do is I review all of the incoming media clips that we get from Ottawa to look at any that are BC-specific. So

that if there's something moving that's here, we have to do on an average day, House cards for Ottawa. Because if the House is sitting, the Minister goes in, she has to have her cards done. And we have about a two-hour window to get them done because of the time difference compared to Ottawa. So that's the first thing we do every morning is to review the clips and see if there are any updated House cards that are needed on media issues that are out there.

The two priorities on a normal day are a daily review of media clips and House cards for the Minister. This orientation of the DG to a constitutive outside audience extends to everyone working under him or her. Ottawa's needs regarding the public, politicized communication of immigration issues and events always trump other components of the workload.

A day becomes a crisis day when you have a bad case that's become all-absorbing or when you have some incident which was unpredicted like the boats, that then requires a focused response. Or when you have something that all of a sudden hits you. Not a bad case, but unexpected: a leak of some nature, some document in your region, or some local issue in your region that affects staff or something, and it's about to blow. . . . It isn't so much a bad day as it becomes a day where you have to give up your agenda, the meetings. Whatever you had going gets canceled, and you become focused with a small group on that specific issue or problem, and you're trying to manage it.

Notably, the orientation of either the crisis-driven day or the day without crisis revolves around communication, responses to the media, and attempts to "manage" the problem.

The theme that everything changes during times of crisis was a recurring one. Crises on the water quickly became crises involving "leaks" to the media and other unexpected releases of information accompanying the unexpected arrival of migrants. This was the DG's generic description of an average crisis day:

Everything changed completely. From the first boat, from its sighting. . . . And I remember it well because on the 19th of July I was at the airport and I was on my way to Ottawa to go to another meeting, and I got called at the airport in the lounge by [a colleague] to say that a boat had been sighted that fit the profile. And my immediate reaction was, "Can you just manage this for a couple of days while I go to Ottawa?" And he said, "I think you need to stay here." And of course, he was right and I did stay, and from that moment on, my complete agenda became nothing but the management of these arrivals and the processing of these migrants, which occupied all of my time and all of my space from six o'clock in the morning sometimes until ten to midnight at night, seven days a week for that whole time period from July the 20th to the end of September. (Interview, Vancouver, August 2000)

The boat arrivals far exceeded the "average crisis" and changed everything in the workplace. An official higher up in the department in Ottawa recalled vividly where he was and what he thought when he saw the first images of the first ship on the water: "This is the beginning" (Interview, Ottawa, March 2001). Like other officials who had been watching marine interdictions on the rise in the United States and Australia, he sensed that this was the beginning of something different for Canada. The sense of crisis began with the first arrival and served as the backdrop to decisions and policies devised in a department that was understaffed and underresourced.

The arrivals proved cathartic for those state workers involved, prompting them to distinguish between a time before and a time after "the boats." For some, this was the most remarkable event of their careers.

In all of my nineteen years of service, I had never lived anything like it. And Frank . . . said in his thirty-two years in immigration, he'd never ever lived through a period like that where there was that kind of focus and pressure and stress in the organization. (Interview, Vancouver, August 2000)

It matters that the boat arrivals were a crisis for people responding; the implications of a state in crisis will be addressed later. Stress levels were high, and people were exhausted. A "beleaguered bureaucracy" (Morris 1985 with reference to the INS) to begin with, CIC was stretched to its limits regionally in terms of both financial and human resources. As a result of the interceptions, people and other resources were diverted, and the department was bifurcated regionally.

The clients still come through the front door; the citizens still get their citizenship. So the rest of the organization has to keep working because the applicants don't stop; the processes don't stop. So you bifurcate your organization. And you have one part of it which is continuing to process as best it can the clients who are coming in through the ports, in the downtown offices, etc., and everybody else in the organization who would have done anything else is completely diverted to assist with the crisis that you're working on. (Interview, Vancouver, August 2000)

This bifurcation extended nationally:

We got some staff that came from back east. We also took some of our own staff—for example, examination officers at ports of entry—and moved them

over to Victoria, and then when they left the ports of entry, we brought back
retired staff and others, casuals and so on, to backfill while they were gone.

Clearly, the disruption to the organization at this time reached far beyond
the sites of interception, processing, and detention, and into the corners
of every office that lost managers, interpreters, and immigration officers.
These statements reveal a bureaucracy caught unprepared and stretched
to its limits.

Being underresourced meant that the contingency of a successful
response rested on the performance, work ethic, and leadership of employees. While people worked tirelessly to coordinate resources and communications behind the scenes in Vancouver, still others worked in pressure-filled
environments on the frontlines at sea and at the Esquimalt naval base,
where migrants were processed. There, managers and immigration officers
felt the impact of slow decisions, as well as the need for resources and for all
manner of collaboration to be unified as quickly and smoothly as possible.
Some of these people would send information and news clippings to senior
managers and leaders in Ottawa, expressing frustration with the lack of
response.

The contingent geographies of human smuggling interceptions at
sea thus often translate into management crises, and crises precipitate
exceptions.

Processing the first two boats of migrants proved particularly challenging. When the first boat was sighted, CIC employees had only twelve hours
to set up facilities at Work Point Barracks at Esquimalt. They devised procedures quickly for processing, adjusting and improving them over time.
In interviews, respondents recounted that time at Esquimalt was marked
not by days but by boat. "The first boat was a really long process. They were
starting to make plans to try out different things, and I helped with that. In
fact, by the time I got there, everybody was bagged" (Interview, Vancouver,
August 2000).

In addition to setting up procedures to provide due process to refugee
claimants, employees attended to health and safety issues.

> These poor people in Victoria just went mental because all of a sudden they
> get a boat, so they'd be phoning around, like ok, I need twenty, and then
> you think of the [motel] rooms. Because you have the officers, you have the
> [Marine Response Team], the translators.
>
> And then accommodations, and then supplies. It was just constant. Like
> [the Department of National Defense] would want [something], and they

needed bras for the ladies, so somebody would have to go out to Kmart to get bras for the ladies. I mean there were just so many things! It was just huge last year. (Interview, Vancouver, August 2000)

Logistics alone overwhelmed those coordinating this work in Victoria and Vancouver. Logistical challenges included finding medical personnel and equipment, such as an x-ray machine.

I remember phoning Health Canada, and they said, If it's a quarantine issue, yeah, I'll talk to you, otherwise talk to provincial health. And I was boggled. I said, I need an x-ray machine! I have to do x-rays of all these people for TB. Where the hell am I going to get an x-ray machine? You can't go buy those things. You can't rent one. That was a big thing. I'm an immigration officer. Where do I get an x-ray machine? (Charlton et al. 2002, 39)

Many employees working in Victoria and Vancouver described the physical and emotional exhaustion brought on by stress and overwork.

It is completely emotionally draining. In the normal working days and weeks when I would go home at 5:30, 6:00 P.M., I could take work home with me in the evening, as I would often do, and work for an hour or more in the evening in preparation for the next day. . . . In the times of boat crisis, when I might have got home at six or seven at night, . . . I was completely unable to look at anything else. I was completely exhausted. I had no energy, I had no desire, I had no care. And so I just stopped doing it. (Interview, Vancouver, August 2000)

This statement from a seasoned manager suggests the extent to which the response stretched and exhausted individual employees and the organization as a whole.

When asked to describe how the boat arrivals affected them personally, some described the stress that being overworked placed on personal relationships. One manager spent so much time worried about a potential leak of information pertaining to the boat arrivals one weekend that his wife confronted him, concerned that there were problems in their relationship that he had been hiding from her. Still others joked about designing T-shirts that said, "Not tonight honey, I have a migrant." With the tremendous demands placed on employees, humor became a coping mechanism.

Communications and "the Need to Know"

All aspects of this logistical crisis were amplified by the constant and conspicuous presence of the media. In interviews, the most frequently cited source of stress in the response to human smuggling was the need to interact with and manage the external environment, constituted not only by smugglers and their clients, but also by other institutions within Canada: other federal departments, provincial ministries, lawyers, advocates, NGOs, and suprastate institutions. The media proved the most stressful presence, and the one most often associated with crisis in interviews. Bureaucrats struggled with the weight of public image in their day-to-day work. While the boat arrivals were an extraordinary event in some ways, they were par for the course in others.

Descriptions of both crisis and noncrisis days revealed that every day began with analysis of media representations. A "crisis" day involved those media stories perceived to be out of control. During the boat arrivals, "communications employees worked only this issue for two months, nothing else" (Interview, Vancouver, August 2000). As events unfolded in the form of a crisis in the media, the media became part of the crisis for the federal government.

Most remarkable ethnographically was the extent to which the bureaucracy's daily operations were oriented to responding to the media. From the office layout to the architecture of daily routines, media representations and public opinions—which were particularly shrill in 1999—made up the everyday life of the communications employees who dealt with them. Many respondents complained of the general media climate and its pervasive impact on their day-to-day work. One long-term civil servant referred to this climate as "gotcha journalism," a shift he associated with the Watergate scandal in the early 1970s that led to Richard Nixon's resignation as United States president in 1974. Respondents explained repeatedly that the media created a climate in which "people in government are not allowed to make mistakes." This reality affected their work to the extent that they anticipated media coverage and the assignment of blame as they would moves in a chess game.

Employees at CIC started every morning reviewing news clippings that included any mention of the department from national and local news outlets the day before. During the boat arrivals, these were compiled by a team of five employees hired by National Headquarters to gather, compile, and circulate news clippings nationally (Interview, Ottawa, March 2001).[8] In Vancouver, these were the subject of running commentary in the lunchroom and across e-mail exchanges about certain journalists who were the least forgiving in their commentaries.

For managers, DGs, and the communications department, studying these morning clippings became not only how they started their days, but sometimes how they spent their whole days:

Your days become patterned because you come in at 6:00 to 6:30 A.M.,[9] and you do clips first. . . . That morning you'd read about what the media said about you yesterday, and you'd then determine were they accurate or not, did they get it right? If they didn't get it right . . . that becomes part of your work for the day.

During "the summer of the boats," the DG of RHQ started every morning reviewing multiple newspapers and daily television news shows and analyzing them with communications employees. A leak or negative portrayal could throw the office into disarray until it was adequately handled. My review of in-house files revealed e-mail exchanges between employees in Vancouver and Ottawa encompassing several days regarding statements in the press.

Work with the media proved both reactive and proactive as bureaucrats simultaneously attempted to predict what journalists would say and do:

Then, we've got some things happening today, we're moving some people, blah blah blah blah. The media will know or want to know. How are we going to manage that? So you set up: Are we going to have a media conference today? If so, who's going to do it? What are our Q & A's? What are we running here? What lines are we putting out? And what are our anticipated questions from the media, and how will we respond? Work that through. So you're reacting to yesterday, you're getting ready for today, because you're doing some things that you want to tell them about. And then you're trying to think about what else do you think they're gonna ask us that isn't necessarily a today issue but might be coming at us from left field. And so you're working through all that in your mind as well.

This quote conveys the extensive work put into dealing with the media in triple time: thinking about what was published yesterday, what might come out today, and what will likely arise tomorrow. Far from extraordinary, however, this relationship with the media was routine. What was extraordinary during the summer of the boats was the quantity of coverage, with some eighty front-page articles, and the seemingly insatiable craving by the media for information on the "sexy" topic of human smuggling.

These anxieties suggest that dire consequences would accompany any mistakes made before the press, a perception that compounded the stress of these interactions.

And then your spokesperson that goes before these cameras that are—some of them are live—we were on live feeds back to CBC Newsworld or CTV News-net.[10] I mean these people are being seen across the country. So it's really important that [they] project images of knowledge, of understanding, and don't get riled, don't get emotional. And we were lucky because we had good spokespersons who projected those kinds of images. And all of that is some luck and a lot of hard work and preparation.

Despite the team's extensive hard work and preparation to avoid betraying emotion, however, the media always managed to throw something unexpected their way.

And usually, every day, there would be one item in the media that would come out, a press story or something that was totally unexpected, unanticipated: something that we couldn't predict. And so when that was happening, you had to have a capacity with the unexpected, the unpredicted, to be able to take it and figure out how to manage it as opposed to take it and fret about it and worry about it and let it eat you up. The need for flexibility, the need for understanding that a lot of what was going on wasn't within our control—would come from outside—was fundamental to knowing: given that reality, you'd better be able to make the time for the unexpected and manage back.

For CIC, human smuggling became synonymous with media excess. This in turn meant that human smuggling was synonymous with a crisis in the bureaucracy, not only because of the logistical and operational work required, but also because of the extraordinary resources invested in working with or around the media.

If the media proved to be a domineering force in office life, this effect was multiplied during field operations. Aerial photographs of the first ship intercepted in July show responders approaching from the platform ship, the Coast Guard's *MV Tanu*, via RCMP Zodiac, as a boat and a floatplane hovered nearby filming and photographing the action. During the interceptions, media outlets competed for coverage. At one point, journalists drew straws for spaces on a shared boat. They competed for satellite service with cell phones, and federal employees were forced to use land lines for secure communications.

Meanwhile, CIC struggled to manage information and promote the image of being in control, although constantly having to respond to the media left bureaucrats feeling powerless. "More than anything else, communications dominated all of it and completely absorbed us each day as to how [to respond]" (Interview, Vancouver, August 2000). Through the media, they strove to convey a sense of power, of having the situation under control. Nonetheless, the government was rarely portrayed by the media as being in control (Hier and Greenberg 2002).

The importance granted to the media manifested in the dominance of the communications branch within CIC, referred to by one person I interviewed as "the tail wagging the dog." Not just anyone in the bureaucracy could respond to the press, and only rarely did those on the frontlines communicate directly with the public. This branch faced a tremendous amount of pressure in Ottawa and especially in BC, where they gave interviews and conducted press conferences for local, national, and international news outlets.

> Every day, that office, which is just across from me here, had two phones in there and they would ring constantly the whole time all day. So [they] had stacked up calls, people to get back to. [They were] getting help from staff, but this was the kind of pressure that [they] had. (Interview, Vancouver, August 2000)

This resulted in a chain of command wherein those responding actually had to run their plans by communications first, a practice that was viewed with a healthy degree of cynicism by those employees who were not in charge. Even though communications employees did not work directly with clients, they were among the first flown to interception sites to begin working with the press.

Employees processing arrivals at Work Point Barracks in Esquimalt also felt intense pressure from the media. Once the migrants entered into processing at the base, the media clamored for coverage and set up camps along the perimeter. One person working on the base described the media presence: "The stupid media . . . they would be camped around the fence at little vantage points. They were filming through the fence. That drove me crazy" (Interview, Vancouver, August 2000). Another person working at Work Point Barracks with many years of experience described this time: "The media scrutiny was something I've never experienced. They were on us like hawks all the time. Vultures. We couldn't walk from the gym to the trailer" (Interview, Vancouver, April 2001).

In interviews, respondents repeatedly brought up the media as being the most stressful aspect of the response to human smuggling. A manager working at Esquimalt said, "Everything you did, you were in the fishbowl. I just waited for something to go left" (Interview, Victoria, March 2001)."Fishbowl" is an apt description of work in immigration during that time: one had the feeling of being constantly watched.[11] The embodied state occupies a fishbowl, simultaneously seen and seeing itself vis-à-vis the media.

There were daily press conferences. Some in CIC were critical of these interactions. In interviews I always asked people what role the media played in the response; verbal answers were often accompanied by powerful body language: sighs, slouches, groans, a roll of the eyes, a shake of the head, and looks of utter frustration. Note the visceral terms with which one employee described the media and its impact:

> We couldn't get out in front of the cameras fast enough. As soon as another issue came up, we were arranging to get in front of the media again. They were all slobbering. They wanted us. They could smell blood. (Interview, Vancouver, April 2001)

This statement critiques not only the media, but also the proactive stance taken by CIC during the response. This remark shows the "chicken and egg" relationship between the government, the media, and the public. The relationship is both antagonistic and symbiotic. CIC employees complained constantly, yet also worked continuously with the media. They complained about the media coverage, but understood implicitly the importance of keeping the public informed and ensuring that the government was portrayed positively. They worked in crisis mode while contributing to the crisis.

Those required to address the media worried about their careers. One manager explained, "You're so tired, and you could say something that would haunt you for the rest of your life. There is always the potential to go left" (Interview, Victoria, March 2001). Mid-level bureaucrats feared that something would go wrong and jeopardize their careers. In the event of a mistake, blame would eventually be assigned. Furthermore, CIC officers often felt scorned in the press for their work, and comments regarding their professionalism turned into derogatory personal remarks. Two exhausted managers addressing a mistake were referred to as "red-faced immigration officers." Accusations during press conferences led to the use of this wording in a headline accompanying employees' photographs on the front page of local newspapers the following day. One brought up this incident in our interview; it was a memory that had not faded.

The media were critical not only of the work at the base in Esquimalt, but of the entire response, from interception to deportation. For some civil servants involved in these events, interactions with the media led to a sobering realization of the close relationship between public policies (such as detention) and the media's authority, as illustrated by the following exchange:

> It was an interesting experience. I think it made me realize how easily the government is swayed by media and public perception. You know, my big thing

was: people would say, "You know, I don't know how you're letting these people in." And I would say, "You know, I've been working with refugees for years at border points and inland. They're coming in every day at our airports, way more than this, and we're not lockin' 'em up."

So it pisses me off a little bit that we're doing all this just because of media and [public] perception. And I think, "Well, look what's coming in at Pearson [Airport in Toronto], look what's coming in at [Vancouver International Airport]! Lock 'em up! Let's be consistent. And I have a problem with that. So all it did—it didn't change me—it just made me realize that government doesn't have any backbone. They don't have any backbone, they really don't.
. . . They're so concerned about public perception.

This narrative links performance directly with perception, emphasizing the frustration with the media's power in influencing the state's response to human smuggling—a response geared to change the image of being "soft" on border enforcement.

More careful journalists pointed out that greater numbers of migrants were smuggled through the airport without being detained (e.g., *Globe & Mail* 2001b). This was met with expressions of cynicism regarding the government's inconsistency. Civil servants themselves pointed out the hypocrisies of policy, as well as the relationship between policy implementation and media portrayals.

A consistent message about human smuggling emerged quickly in the mainstream media: this was the work of transnational organized crime. Beneath the surface, however, this story fell apart as civil servants struggled to assemble a clearer understanding. Decisions about distinct migrations do not take place in a vacuum. Human agency, public policy, public opinion, and public discourse all collide in the field of immigration.

The "fishbowl" climate perpetuated a sense of fear among bureaucrats, which was an interesting contrast to the image of the immigration officer as a figure of power. Most expressed feelings of powerlessness and the frustration "that government doesn't have any backbone." With federal departments negotiating for power via the media, individual civil servants are left vulnerable to the assignment of blame as they contend with the design and implementation of "policy on the fly."

Policy on the Fly

A small group of people did the work of many, and the command center in Vancouver illustrated this. While interceptions were being made in the waters off Vancouver Island and the more remote Queen Charlotte Islands, hours away from downtown Vancouver by plane, RHQ served as the central base of command and primary node of communications. Like the Wizard of Oz, the moniker "Command Center" gave collaborating agencies the impression of a sophisticated hub of operations. This was a running joke to those within CIC who knew that Central Command consisted of one man working in one of many small cubicles at the office, rigged with the same computer and telephone as the rest. The major difference was that a foldable cot was moved in for him to sleep on during nights when an interception was under way, and he had to stay at the office to receive calls from BC and Ottawa. Often his was the only desk light glowing well into the night. Co-workers would bring him breakfast when they returned the next morning. This reality highlights the demands placed on a small, underresourced department. It also underscores the contingent nature of policy on human smuggling.

> The Minister was so public, as she has been, a number of times that she's spoken on it. She's said, "We have to focus on the smugglers. We have to stop it where it starts over there. If it doesn't, if it gets here, we have to catch them, we have to try them. . . . We have to create the disincentive." We have to wherever possible detain when it makes sense and remove when people have exhausted all processes. We will give all due processes. We have the [Canadian Charter of Rights and Freedoms]. They have rights. They go through the refugee process and they have all the appeal processes, but then when they've exhausted processes, they must leave. They must go back. And that's what we have done here. And we have been, in British Columbia, the cutting edge of all of it for more than a year now, with the two largest removals in the history of immigration in these past few months. We are witnessing things that have never happened before. And so, because they are new, we don't have a long historical experience to base what we're doing on. We just have current knowledge and what we think will work and ought to work. (Interview, Vancouver, August 2000)

One of the reasons why there was so much stress in 1999 is that this movement was unprecedented in BC's recent history. Despite the simulation exercises months earlier, there was no clear policy driving a response to marine arrivals, which resulted in the invention of "policy on the fly," or as one respondent put it, "what to do to not get the government in trouble" (Interview, Vancouver, July 2001). [12]

> A lot of what we were doing here—because it was so new—we were in fact doing it here and we were almost in a way making policy here. Because we

didn't have guidance in some of the areas, we had to try and figure it out and do it ourselves. (Interview, Vancouver, August 2000)

The most controversial contingent policy designed "on the fly" was detention.

Detention as Containment Strategy

The fact that procedures and policies were designed *after* the occurrence of major events supports the notion that policy represents as much a mirror of the past as it does a guide for the future. In 1999 there simply was no guide to handling marine arrivals, and bureaucrats had to draw on their creative resources to manage the crisis. In the long term, a common element of policy on the fly was detention, one of several strategies outlined in chapter 2. Since Canada does not detain refugee claimants as readily as the United States and Australia do, and because detention is costly, this was a controversial decision. Employees indicated that there was not much discussion at the time about why—or whether—to detain. Many spoke, however, of the immense pressure from the media and the public to "do something." One respondent linked policy on the fly to the public outcry over the arrivals:

So it was this big outcry, so they locked them up. Yes, in the end, that's probably what helped deter the boats. But initially, I bet you, if people were honest, that wasn't why they locked them up. It was just like, we've got to do something, and they were just bending to public pressure to keep them locked up. Because if they were really concerned. . . . That's why it was good, in a sense, for me, when [one journalist] started to say, "Wait, hang on, look at the number of [Chinese nationals] coming through [Vancouver International Airport], why are the two different?" And [the government] sort of said, "Well, this is organized smuggling." . . . This might be a bigger operation, but one guy bringing in four people is no different than this person bringing in 150. Because this guy does it ten, fifteen times a year. (Interview, Vancouver, August 2000)

She suggested that the decision to detain was at least partially wrapped up in the importance of image, the visibility of the arrivals, their association with organized crime, and the desire to sustain the government's integrity in light of the criticism of its response (see Hier and Greenberg 2002). While detention is among the more expensive short-term solutions to human smuggling, it is a visible expression of strong governmental effort to contain the problem.

There was, therefore, a direct link between the geography of human smuggling and the policies designed and implemented in response. Boat

Figure 7. Local and national newspapers frequently published photographs of minors who were refugee claimants, here shown in handcuffs in the custody of immigration authorities as they moved in and out of institutional facilities for Immigration and Refugee Board hearings. Faces of claimants have been altered to protect their identities. Published in the *Vancouver Sun*, 20 August 1999. (Courtesy of Asian Studies Library, University of British Columbia.)

arrivals were construed as a crisis in the media, resulting in detention as a strategy to contain migrants and bolster public image. Similar to many trends in the "regionalization" of displacement—the containment and "protection" of the displaced in their regions of origin—recent strategies are intended to reduce the number of asylum seekers arriving "spontaneously" and to combat the image of "weak" or "soft" border enforcement. Both strategies attempt to contain disorder by moving those displaced farther away from sovereign territory of more powerful states.

News images that portrayed a government not in control of its borders soon gave way to images such as that in Figure 7, showing handcuffed claimants being moved in and out of detention centers.

After the first boat was released, many immigrants did not appear for their claimant hearings; these individuals were assumed to have reestablished contact with smugglers and gone on to the United States. The reasons this group was treated differently through detention en masse went largely unexplained. Officials cited the involvement of organized crime, the failure of claimants to provide identity documents, and the fact that they were likely to flee. The same could be said, however, of many arriving at airports and land borders who were released during the same time period. People provided different answers to explain this in interviews. One official in Ottawa noted that the "boat people" were a "small problem," relatively speaking, but that the phenomenon of the arrivals "struck a nerve" for three reasons: "It struck a nerve because it was first of all a direct attack on Canadian sovereignty. Also because they arrived illegally, not intending to claim refugee status. And because they were clearly not refugees" (Interview, Ottawa, March 2001). His remarks echo those of many others who brought up illegality and the belief that the refugee program was being abused by people assumed not to be Convention refugees. Another official based in Ottawa offered additional reasons, again making reference to erosion of the integrity of sovereignty, the law, and the refugee program:

> Why does a small number invoke such a large reaction? . . . Because you have to situate it within the context . . . of the public perception and attitudes towards immigration and public perception and attitudes about an immigration system that is out of control, that doesn't have proper enforcement, that doesn't have integrity for refugees applying and working through the system. . . . And so you just put into it . . . these *very visible, symbolic boats*, which spark reactions unlike the airplanes that are landing as we speak right now in airports across Canada that have on them refugee claimants. Because they are *so* symbolic . . . these pictures that you can never forget, seeing these rust buckets

with people in these absolutely abysmal conditions and what they've put up with, and what they've suffered for, and how it's all part of this large organized crime, and how it's endangering their lives. It's a human interest story.

And then you have this. Which is not only the human interest drama of people risking their lives to get here and of organized crime working its tentacles in the worst way, but you have the two key factors which enraged Canadians and British Columbians about these people, these Chinese migrants. . . . So not only was it surreptitious entry which was being desired—not being seen, not being caught—but secondly, they had no desire to be in Canada. They were on their way somewhere. They were going somewhere. This totally enraged Canadians. . . . That's why a small number has huge impact.

Asylum seekers who arrive by boat are positioned as illegal and disorderly. In Tim Cresswell's (1997, 2006) work, these bodies leaking out of place are least desirable and are punished in moral terms for their disorderly essence. They are more likely to be racialized, criminalized, and segregated from others geographically (Sibley 1995). Although bureaucrats often posited human smuggling as a challenge to Canadian values, this position resonates with other immigrant-receiving governments that wish to efficiently manage transnational migration.

Because it hits at the core of what Canadians have determined to value from their programs in immigration and citizenship. We want to bring people to come to Canada for family reasons, for refugee reasons, or for economic reasons. We will protect people who get here and make a claim, and if they're refugees, we want them to stay here and be good citizens. This small number of people didn't fit it, didn't fit the images upon which we've built the traditions, the values of immigration. (Interview, Vancouver, August 2000)

Because spontaneous arrivals threaten state sovereignty, the response is to exercise border enforcement as an expression of sovereignty. In fact, Operation Osprey, as the response was called, was designed "to enhance immigration's profile on the water front" (e-mail from CIC employee, 4 January 2000, released to the public).

Another respondent connected the public response to racialization. She associated "normal" physical appearance with those around whom the public rallies support, whereas those racialized as "other" than the norm did not receive support.

The people feel they're getting screwed in a sense, like here's all these people riding on our shores. But I can't help wonder if it wasn't a boatload from England what they would have done. I think it's important to look at that. . . . And then you get [these arrivals], and everybody's saying, "Send the bastards back." I don't know, I never see them rallying around a family of . . . I don't know, I mean that's just my perception. If they look normal, and everybody says, "Oh, immigration's so bad, they should just let them stay. Look at the big bullies picking them up." But they want the Hondurans gone. You don't seem them rallying around them. You don't see them rallying around the Chinese. . . . I wonder myself. If you got a boatload from England, what people would say. I really do, I think it would be different. (Interview, Vancouver, August 2000)[13]

By suggesting that the public rallies around people who "look normal" but wishes enforcement against people of color, this respondent associated "looking normal" with whiteness and called upon racialized images of settlement and exclusion in Canada (see Anderson 1991; Bhandar 2008).

Another respondent provided an additional bit of context surrounding the ways in which this movement was dramatized by the media and the public, suggesting that the response resulted from public opinion and media representations. Given the prominence of the boat arrivals from Fujian, most members of the public would probably guess that China was the largest source country of asylum seekers in British Columbia in 1999. This was not the case. Few realize that Mexican nationals actually composed the largest—if not necessarily the most successful—group of claimants in BC in 1999.

Our number one source country was Mexico. And that would drive me crazy. . . . There's other areas that we should be looking at too but nobody seems to think so. But Mexicans come in. They don't cause a bother, they come in in little trickles, even though the numbers are high, and nobody wants to look at it. And it just kills me. . . . So every time I'd be in a meeting with Ottawa and whatnot, and of course the Hondurans were the big thing. Now granted, the Hondurans are not the nicest people in the world. But again, it was the media. And I'd be in the meeting going, "You know, we only get a few of those. I know they're causing problems. You want to look at the Mexicans?" Everyone ignored me. (Interview, Vancouver, August 2000)

As this scenario indicates, the boat arrivals and the response were never separate from the national political stage. When I began my research during the summer of 2000, I was told that people in the BC region were being "watched like hawks" by Ottawa because of the upcoming election (Interview, Vancouver, June 2000). Following rumors of boat sightings in the ensuing months (known as "false alarms" at RHQ), the DG in Vancouver mentioned receiving direct calls from the prime minister's office to find out what was happening.

This issue was a crisis not because of the number of refugee claimants, but because of the method, geography, and visibility of their arrival and the accompanying political implications. Crisis begets policy, but the response did not fix policies already in place. An absence of policy means an absence of decisionmaking and public consultation. Without a set policy, the connection to politics implied by respondents could be actualized. And the result of a failure to set policy resulted in crisis. Those bureaucrats most affected by this reality in 1999 were the managers in BC who were responsible for an effective response but lacked the policy and resources necessary to guide them.

Because of the degree to which the daily operations of the bureaucracy were geared to interface with the media, there was no action outside of a performative mode of being. The bureaucracy responded to depictions of disorderly migration and excelled at the performance of crisis, attempting to create order out of chaos.

Bureaucracy as Communication Network

Regional employees blamed the tension on the hierarchy, and they resented the inability of bureaucrats in Ottawa to understand the geographies of interception on the west coast and the nature of other pressures they had to negotiate locally. Such misunderstandings were affirmed by a report that people arriving from Ottawa for meetings asked regional employees to arrange for them to visit the coastal interception sites over lunch. Exasperated, the recipient of this request explained the logistical impossibility of executing such a trip during a lunch hour, given the distance to the sites and the inaccessibility of the coast. Such a visit would require a day-long excursion by air.

During the summer of the boats, when Ottawa waffled on decisions, CIC employees in Vancouver often explained it as resulting from misunderstandings due to differences in their respective places in the hierarchy. Respondents mentioned, "For them, this is a career. For me, it's a job." This difference in perspective was evident in the employees' frustration with the way government policy was tied to public opinion, as demonstrated in the inconsistency with which it approached detention by sea versus by land. This reinforced the conviction that responding to human smuggling was a job for some, and a political issue and career trajectory for others. Still, CIC in Ottawa held the "monopoly on vision," in the words of one respondent.

During the height of the response, intense daily conference calls were made in an attempt to bridge the distance between RHQ and NHQ. These conversations included a large number of people in both places, including a representative from the Minister's office, deputy and assistant deputy ministers, lawyers from legal services, and people working in communications, refugee issues, enforcement, and operations on both ends. In the beginning, two daily conference calls took place between RHQ and NHQ. These became part of the daily routine for employees in Vancouver and Ottawa.

For people in RHQ, the stress of these calls with Ottawa added to the larger burden of keeping everyone informed of new developments. They attempted to separate out "need to know" cases and prioritize them. "You had all your calls to your partners who you have to keep in constant contact with: RCMP, National Defence, you've got all that stuff going on. You have constant, constant media pressure" (Interview, Vancouver, August 2000). Communications to employees at RHQ suffered as a result.

> It's really tough. . . . I mean we pride ourselves in the organization in being able to inform our staff about issues in the organization that are going to become media events before they are published in the media. Because we always want them to hear about it from us first and not read about it in *The Globe & Mail*. But the reality on this one was that we were so, so absorbed in it and having to manage it that we didn't have the capacity to continually keep our staff updated on what was going on. And because the media was so, so absorbed in it and so reporting on it morning, noon, and night, we couldn't possibly keep up with the media. And so the staff was hearing about it from the media before they got it from us, and there's nothing we could do about that except hope that they would understand what we were living. (Interview, Vancouver, August 2000)

During stressful times, disputes erupted regarding who "needed to know," and long-standing antagonisms between employees in different branches then played out. People working in communications often found themselves at the center of such disputes, because they were determined by the department to be those with "the most" need to know. They therefore held significant power in determining with whom information would be shared. Tensions ran high during conference calls when leaders, managers, operational people, and lawyers would come together across time and space to make important decisions. They could feel the distance and difference separating them, which is best represented by some respondents' belief that the people on the telephone in Ottawa had never worked in the field.

Infused into these logistical operations was the demand for the circulation of information among people geographically far-flung. At these times, the informal network of state practices became more formalized to connect people across time and space.[14] In addition to the telephone, which enabled instant communication and large meetings via conference calls, another mechanism for the circulation of information that radically altered the time–space relations of government personnel is e-mail. The immediacy of e-mail created additional expectations of a speedy response from Ottawa and rendered memos redundant.

One DG described the amount of time-sensitive work generated by e-mail:

> On an average day, my computer will receive from 100 to 200 e-mails. So you have to figure out a way—because you don't want to become a bottleneck—of keeping your system cleared. . . . So a key objective for this position is by the end of the day . . . that your e-mails have all been checked and moved out to people that need to have them or trashed if they're not of use. And that requires the strong assistance of an executive assistant who is trained and knows what's important in your absence and knows what needs to be routed quickly if it's of an urgent nature. So e-mail has complicated what is already a very complex job. . . . And of 100 e-mails on an average day, there may be five or ten that are truly important to you, and the rest aren't. But you'd better make sure that those five or ten are ones that you or somebody else has looked at and has moved because the world of e-mail is the expectation of the world of quick response and if you're not there, you could miss out on something that could be significant. So that's become brutally difficult. (Interview, Vancouver, August 2000)

E-mail also became a central part of record-keeping in the bureaucracy. Many of the files that I reviewed were e-mail exchanges between Vancouver and Ottawa that documented the drama as it unfolded in real time with each boat. The quotidian nature of these exchanges makes them an important historical record, as well as a tool for analyzing ethnographies of the state.

A sea of files of other types documented the arrivals, including instructional manuals and policy interpretations; legal opinions; e-mail messages, letters, memos, and reports; communications strategies; speeches; binders of photographs; results of research into infrastructure and logistics such as interpretation and security; reports on interceptions at sea in other places; contingency plans; and all manner of records on the ships themselves.

While there is an entire industry established within CIC to monitor, destroy, and "vet" those documents released to the public, one result of the significant downsizing of the federal government in the 1990s has been a reduction in the number of people in clerical and lower-level administrative positions. This human resources issue is reflected in the poor record-keeping capacity of the department and in the deficiencies in what bureaucrats told me was once a more robust archival system. Mid-level bureaucrats now must largely decide themselves what to keep and what to discard, *and* they are responsible for doing so themselves. Some complained about not having an administrative assistant to organize and maintain the files. Those involved in the marine response who discerned the importance of this work in the historical record had the foresight to print and file many of their e-mails. Some also decided, with the support of their superiors, to write reports on the events; others at the center of the response were less interested in writing or simply did not have time to do it. This shows the role of human agency even in record-keeping within the state, and the reliance of history upon those who see the big picture and wish to contribute to the records. This limited not only the records kept on the arrivals from a historical perspective, but also *who* could be informed and in what way at the time.

Because records on the marine arrivals were maintained separately within different divisions such as communications, policy, and intelligence, few people, if any, knew everything on record. This contributed to a dynamic wherein some people worried that I had been given access to a "smoking gun" of some sort, whereas others more familiar with the files insisted that there were no hidden secrets (Mountz 2007). The confusion surrounding security-related information contributed to the aura of intrigue, allowing a mundane issue such as irregular migration to become a crisis.

The subjectivity involved in creating any historical record contributed to the short institutional memory regarding human smuggling.

Of course, bureaucrats and researchers were not the only people interested in the records on the boat arrivals. A steady stream of requests for information flowed into CIC, enabled by Canada's Access to Information Act. Once again, these requests, made by lawyers and the media, and the bureaucrats' response to them gave me the impression that there was far more subjectivity involved than I would have anticipated.

The steady stream of access requests created an enormous amount of work for people in CIC, and largely shaped their daily schedule. At times, employees invest a large percentage of their workday filling these requests. Given the amount of paperwork that they handle on a given day, any information request has the potential to overwhelm. One person joked that he was waiting for just a few more access requests to come in; he had calculated that he could then cease his regular work assignments and spend his remaining working years fulfilling access requests. The effect of these requests was that people in CIC actually had to think through the ways in which they documented things *before* they were requested. They even shaped the wording they used when communicating with the public, so as not to divulge certain pieces of information that could set off a stream of access requests on a given topic. In the relationship between the department and the media, bureaucrats were always aware of the possibility—indeed, the probability—of access requests, and therefore were proactive in the internal documentation of information as it related to an external audience. Some respondents recounted their efforts to control information by erasing e-mail messages before they could be backed up by the server and thereby imprinted in the institutional record at the end of the day.

So while, in theory, access requests enabled the public to be informed about the government, in practice they also served as a barrier to communication between government and civil society; an imperfect way to keep the public both informed and distanced through time and the protection of information.

Contingency, Fault Lines, and the Wait

There was a long-standing division within the department between the work of enforcement and that of facilitation, referred to as "the division between church and state." Given the reputation of "softness" that immigration officers perceived they held within the "federal family," those working on enforcement within CIC felt particularly frustrated by their status in the department. Planning for human smuggling, clearly positioned within CIC as an enforcement issue, fell prey to a lack of a long-term commitment to enforcement by the department's leaders.

Nation building is intimately bound to the policing of international borders (cf. Hage 1998; Nelson 1999; Razack 1999; Sharma 2001; Nevins 2002). The boat arrivals raised important questions regarding Canada's international role as a humanitarian, refugee status–granting nation of immigrants versus one that is "too soft." The migrants from China were actually en route to work in the United States as undocumented laborers, like the other estimated 12 million who work there for low wages, in poor conditions, without access to social benefits. This posed challenging questions regarding Canada's role as a transit country.

There was also a philosophical divide over the distinct mandates of CIC as facilitator of legal immigration versus enforcer. People within CIC disagreed about whether they should be involved with boat arrivals, and if so how, and they held distinct positions depending on the framework in which they worked. Those whose responsibility was to project a public image used smoke and mirrors to create the impression of an organized, meaningful, and unified response to international organized crime. Their message was that the Canadian state is not vulnerable. But those within operations and planning did feel vulnerable and unprepared as they clamored for more resources to do their work and to learn more about smuggling networks. Those working in enforcement felt that they represented the "dark side of immigration . . . the things that no one wants to talk about" (Interview, Victoria, March 2001). They complained of a lack of support from the department in the areas of policy, finance, human resources, and general respect. When they made recommendations to prepare for smuggling movements, they felt ignored. Their interpretation was that CIC politicians did not like the image of immigration officers working out on the water as enforcers, removing people from, rather than attracting them to, Canada.

The difficulty of planning and preparing, however, was not only a function of mandates and philosophical divides within the department, but was also a budgetary dilemma. Government budgets are rarely designed to provide resources needed for *possible* costs.

> Last summer, we stopped at four. We thought we could have got two, three, or four more. If we'd have got that many more, it would have become an actual crisis, not just a BC-run, managed-well crisis. And why? Because by the end of the fourth boat, we had reached saturation. Our capacities had all been reached. We had no more detention space. Our staff were getting exhausted. And if we had somehow got more, it would have become a national crisis. The responses would have been different. We might have had army camps opening up; I don't know what. And there would have been a lot more presence of military than there was now because we didn't have the people. We'd reached our limit. (Interview, Vancouver, August 2000)

The bureaucracy's resources shaped its response to human smuggling. When boats stopped coming, resources dwindled.

The challenges to planning for the future were not only financial but also logistical. Planners invested funds in reserving human resources such as medical personnel, security, and interpreters, as well as emergency supplies. CIC entered a number of short-term contracts with a variety of detention facilities in BC. These half-measures conveyed not only concerns about resources, but also the deep divide about whether and how to increase capacity to respond to future smuggling movements. But smuggling continues regardless, and governments need comprehensive plans for appropriate responses. Better planning might prevent new crises.

Embodiment and Proximity

In his exploration of the "thought worlds" of immigration officers, Josiah Heyman (1995) found that immigration officers were influenced by personal history prior to their entrance to the bureaucracy, and then socialized to a new set of worldviews within the INS. Noting that "social relationships are produced through the bureaucracy" (1995, 263), he argued that immigration officers embody the sovereignty of the state and thus make decisions according to their overarching ideas of self and other that manifest at the border. In CIC, as employees explained their day-to-day work in relation to the mandate of the department, the inconsistencies in their objectives illustrate that conflicting perspectives existed not only *among* government bodies, but also *within* them.

Civil servants are organized into a bureaucracy designed to function despite conflict, across regional and local geographies, with different economic, political, and demographic characteristics. Higher up in the bureaucracy, there emerged a narrative about international organized crime networks facilitating human smuggling, and about sovereignty and protecting the integrity of borders in relation to fighting these forces. Officials discussed the detention of smuggled migrants as imperative (Interviews, Ottawa, March 2001). Bureaucrats in office buildings were far removed from the "frontline work" along borders, at airports, or in inland claim centers. It was easier for these officials to distance themselves from the migrants as individuals and instead characterize them as criminals whom they would work to repatriate. The more removed bureaucrats produced cleaner, more simplistic narratives of human smuggling as "bad," enabled by distance and "dehumanization" (Heyman 1995).

According to Heyman, the bureaucracy must uphold coherent narratives and implement them despite inconsistencies (1995, 277). The strategy of embodying individual narratives not only conflicts with public images of the unitary "state," but also connects authorities to migrants more closely. One federal employee who worked closely with the migrants had herself been smuggled from China to Canada as a child. She described initial interactions with the migrants as "very painful," particularly in relation to the women and children from the boats, with whom she empathized. She was especially upset over her interactions with a young woman who she believed had been raped and wounded during the journey (Interview, Vancouver, March 2001). The bureaucrat reported that many memories returned to her as she worked with the migrants. When she tried to talk to a friend about it, her friend reminded her that she was just doing her job and to forget about it (Interview, Victoria, March 2001). To me, however, she described intense politically charged and emotional experiences connected to her own history of being smuggled into Canada.

Narratives about the response to the boats differed according to the locations, roles, and identities of employees within the bureaucracy, and to their own histories and social embeddedness in their work. These factors influenced how they related to smuggled migrants and enacted policy. Those with closer personal involvement produced narratives infused with emotion, passion, and complexity sparked by intimacy. Likewise, provincial workers in prisons and social workers in group homes came to know the migrants and were upset by their removal from Canada (Interview, Victoria, September 2001). Another CIC employee working in enforcement mentioned spending many hours on a plane with one of the large groups of migrants deported back to China in the spring of 2000. During the long trip back, he came to know some of the migrants and to regret that they were forced to make the trip in handcuffs, realizing how poorly they might be treated once back in China and the overall gravity of their repatriation. These employees experienced points of identification rather than distance and abstraction. Through personal processing of their experiences with the migrants, they came to question the simplified narratives of the "bogus" claimant. One person described "attachment" as a problem in detention centers, where officers and migrants got to know each other. He noted that repatriation would be more emotionally difficult for these migrants (Interview, Vancouver, April 2000). But these experiences opened up a dialogue in that they challenged the perspectives of those who implement immigration policy and caused them to reflect on their day-to-day work.

Ethnographic research shed light on the roles of identity, personality, emotion, and conflict in the enactment of policies where these tensions existed. An embodied state is less of an omniscient abstraction and is more human. This vulnerability of being human relates as much to the media as to the smuggling of people. Those who work in the name of the state come to see and enact the state recursively through media representations.

Subjectivity and the State

Some border crossings are presented to the bureaucracy and to the public as crises, others as routine. As the state "thinks the subject" (Lubiano 1996, 65) of the transnational migrant, it does so in relation to how it imagines itself in the world. Much of the debate about sovereignty and the integrity of Canada's borders demonstrates insecurity in relation to its more powerful neighbor to the south, as detailed in chapter 1. The boat arrivals stimulated an intense moment of nation building, and the public outcry engendered an unusual show of force.

Nelson's (1999) concept of the state as constituting, and as constituted by, others through identity is helpful to understanding the social and political contexts of public policy. The government embarked on a forceful response to human smuggling that was enabled by discursive practices of identity construction. The response to the boat arrivals was of symbolic importance to the Canadian international image. Rather than emphasize its traditional role as humanitarian, the government instead responded with an enforcement stance. Popular and legislative discourse regarding immigration shifted to the strict establishment of legitimate versus illegitimate means of movement. Identifiers of smuggled migrants were wrapped up in moralistic discourse with terminology of "bogus refugees," "boat people," and "illegal aliens." The Canadian response, including the largest group detentions and deportations in recent history, engaged this discourse and reconfigured the role of the state in relation to the production of national and transnational identities.

Sometimes these identities were built on precious little information, as the respondent who discussed transnational organized crime as a "smoking gun" illustrates (chapter 2). This respondent knew what feminist scholars and oral historians have long claimed: that there is no underlying truth to be discovered in interviews, only a series of narratives that people tell, performances offered at specific moments for specific reasons. These are the clues to understanding the performative state through participant observation.

Those enacting policies on the ground struggled with inconsistencies, recognizing that human migration and individual histories were more complex than public narratives. Ethnographies that embody the state are uniquely able to illustrate the fluid array of subjectivities and discourses through which states are constituted (e.g., Gupta 1995). Data presented here illuminate the uneven aspects of state practices where workers embody inherently contradictory sets of ideas.

Ethnographic data also demonstrate the performative nature of human behavior and border enforcement, depicting state practices enacted by people who feel powerless but must appear powerful. When investigated as an everyday enterprise, "the state" emerges as a rather haphazard constellation of actors sharing information and strategies, while operating largely in the dark. As a result, civil servants traffic in information, whether at international airports abroad or via e-mail between regional ports of entry and national headquarters. These dynamic networks prove essential to understanding the shifting spatialities of border governance.

The performative state emerges, exacerbated in times of crisis, simultaneously promoting and succumbing to powerful narratives designed in times of fear and uncertainty. After investigating why and how civil servants enacted the state, I understood them to occupy the gap between the self-conscious performance of a state in control and the more frenetic reality of daily work in a field that is dynamic and uncertain—respectively, Goffman's front- and back-stage performances.

Ethnographic analysis underscores the importance of qualitative research with policymakers, illustrating that the government response to the arrivals depended as much on personality as on policy. Analysis of the state as a series of cultural practices shows that policies are enacted amid conflict, but high-level bureaucrats and a powerful communications branch construct clean narratives for the public. This chapter shows the prevalence of subjectivity involved in state practices, from recordkeeping and the fulfillment of access requests, to the design and implementation of policy. Narratives of the response to smuggling reveal conflicting interpretations of policy that played out across space and time.

The nation-state emerges not as monolithic, but dynamic. However, the desire to present a unified message to the public leads high-level bureaucrats and a powerful communications branch to construct clear, coherent narratives for the public. According to Bakhtinian logic (see Bakhtin 1981; Holquist 2002), which is a monologic narrative wherein one truth prevails, a simple tale emerged: transnational organized crime brought economic migrants to Canada. These were the identities into which migrants were scripted, with legal ramifications as outlined in chapter 4. Individuals working in distinct locations within the state, however, often work in tension with others in ways that disrupt the cleaner, dominant narratives of the state. I moved up and down the hierarchy to interview not only office workers, but also those working directly with migrants. Employees lower down in CIC suggested that coherent, publicized narratives were shot through with ironies and inconsistencies.

Federal responses to human smuggling often involve the performative state flexing its enforcement muscle in a display of power at the border. There is a disjuncture, however, between the idea of the powerful state and the vulnerable state articulated by bureaucrats who felt in the dark, powerless, unprepared, unsupported, and cynical. Some argued that there was never any discussion about why they should suddenly shift policy to wide-scale detention of claimants. Others questioned whether "transnational organized crime" was involved at all. These disjunctures can feed crisis. Dissection of these conflicting viewpoints within the bureaucracy offers the opportunity to recover alternative perspectives, responses, and philosophies beyond those enacted and presented to the public. The disruption of narratives opens the political possibility and space for the kind of social change Nevins hopes to see in civic engagement with border enforcement (2002).

Ethnography enabled closer examination of points of identification, intimacy, and difference through which the state is constituted. The more personal narratives of the work of governmental and nongovernmental employees disrupted the cleaner national narratives of the Canadian self and other, and offer the possibility to reconfigure those relationships.

Crisis: Securitization over Protection

Embodied, states are constituted by individuals, groups, and networks across which information flows, agendas and policies are formulated, and all sorts of follies and foibles transpire. Precisely because they live and breathe, states are funny, emotional, confusing, violent, contradictory entities. So too are the tales told to us by the state, by those who shape its communications strategies and talking points, those "street-level" bureaucrats who implement its policies, and those who police its borders.

The institutional landscape of transnational migration arranges itself along multiple axes of power, from office architecture to the international diplomatic efforts in which bureaucrats engage on a daily basis. The bureaucracy is a key site where information and power circulate, representations are crafted, and identities are scripted. As such, it is a key organizational node through which to understand the shifting spatiality of governance, and through which to question the ontological underpinnings of "the state" by exploring the work of civil servants. During crises and their aftermath, some agendas move forward while others are suppressed. Some institutional actors claim authority to trump others. In so doing, they devise exclusionary practices.

Negotiations of the "long tunnel" that occur beyond and between the lines of policy often tell more about the operation of the nation-state than does analysis of policy and institutional structures.

The administrative function of the modern state has always entailed governance of the daily activities of citizens. As people move at a faster pace in a globalized world, states have intensified their efforts to govern this movement (e.g., Amoore 2007). The state works as an assemblage of institutions, networks, administrative functions, and people organized into bureaucracies.

Bureaucrats make powerful decisions that have permanent material effects on people's daily lives. Yet many of the bureaucrats interviewed expressed powerlessness with regard to decisionmaking. As Herbert (1997, 163) found in his work with police officers in Los Angeles, civil servants attempt to assert their own discretion while constantly remaining aware of the constraints or "normative orders" that structure their behavior.

This chapter advanced the argument of the book by revealing the character and conditions of the state in crisis. It is important to ask *why* human smuggling played out as a crisis rather than as an ongoing global phenomenon for which immigration departments should prepare. The sources of the crisis were many. The distinction between what bureaucrats *do* and *say* was key. They were forced to articulate power while feeling powerless, providing the illusion of order and control while lacking adequate resources and policies to support their response.

As a result, the movements of "spontaneous" asylum seekers or refugee claimants, in spite of their decreasing numbers and the increasingly aggressive legal practices to exclude them, often will translate into crisis. The state is performative, and it excels at the performance of crisis in border policing. Such performances increase the vulnerability of those in search of protection, as they often create situations when enough political will amasses to move policy agendas forward. The more visible the arrival of asylum seekers, the greater the crisis, and the greater its potential to advance policy in the name of securitization over protection.

The ethnography illustrates the extent to which the media generated more of a crisis in the eyes of bureaucrats than did the smugglers. Why this attentiveness to the media, and how did it manage to trump all other issues? CIC's priorities relate to political contexts and Canadian public opinion as mediated by the press. Shifts in CIC priorities caused resistance on the part of federal leaders to the proposition to commit resources to further planning and preparation and caused considerable tension among the ranks.

Hollowed out, the neoliberal state relies on others to do the work of refugee resettlement and policy implementation, having devolved power to collaborative "partners," willing or not. As neoliberal states devolve and privatize, other institutional actors become more involved in the implementation of policies. Various institutional actors both within and outside

CIC took part in the response, supporting the notion that "the state" has permeable boundaries. Some work for other levels of government, while others are funded by government but work for nongovernmental organizations, such as immigration service organizations, immigration lawyers, suprastate institutions, refugee advocacy groups, and even members of the media. They assisted claimants released from detention, represented them in the legal process, monitored their detention, lobbied on their behalf, and told their stories. The failure to plan impacted not only CIC employees, but these "partners" as well, which added to the stress of the state in crisis.

Chapter 4 looks at the work of the media in portraying and even influencing the work of bureaucrats and the efforts of refugee claimants to access the refugee determination process.

Chapter 4   Crisis and the Making of the
Bogus Refugee

ON THE SIXTEENTH FLOOR of the towering Library Square office building in downtown Vancouver sit two senior male members of the Immigration and Refugee Board (IRB); a refugee claimant in her twenties from the second boat intercepted from Fujian; her legal counsel; an interpreter; the Refugee Protection Officer; a representative for the Minister of Citizenship and Immigration; the BC corrections officer who accompanied the claimant, the latter in handcuffs and green prison uniform only moments ago; and along with myself, in the back of the small room, a representative from the United Nations High Commissioner for Refugees (UNHCR). We are all present to witness this claimant's moment of truth in Canada. The IRB members welcome the UNHCR representative deferentially and proceed to run the hearing diplomatically, expeditiously, and humanely, with breaks when needed, such as when the claimant breaks down in tears.

Large stacks of paper are wheeled in and spread before those involved in the case. The claimant's legal counsel and the minister's representative interact with an antagonism born of familiarity. The UNHCR representative attempts neutrality, and I observe quietly from the corner, watching this woman sitting before me, close to my own age, alternately shaking with fear and breaking down in tears. Legal counsel for the claimant do most of the speaking, and the interpreter works hardest of all, although I am aware that entire exchanges, such as legal debates, go uninterpreted.

Invitations to the claimant to speak solicit biographical information and dwell especially on involvement with birth control practices. Conversation moves from the removal of intrauterine devices to sterilization to abortion to an abusive husband. The claimant has one child and has spent the last year in detention in Canada. My thoughts turn to the involvement of governments in the intimacies of our daily lives.

Timing appears to be everything today, from the fact that this claimant was initially issued an exclusion order by an immigration officer after arriving on the second boat and later invited to make a claim, to the complaints of her counsel that CIC failed to respond to repeated requests to release information in her file in a timely and comprehensive manner (he notes sixty pages missing).

These have become important legal issues with ramifications for claims from others who arrived by boat, and the lawyers debate the time when processing ended and detention began at Esquimalt and whether or not this claimant—along with others—should have had access to legal counsel sooner. At stake is the inclusion of notes in the claimant's hearing from the early interviews during processing at Esquimalt, before she had met with legal counsel. Today, her legal counsel requests that the board reject everything included in her personal file—including these notes—unless the entire file is disclosed to him in a timely manner by CIC. CIC lawyers argue that they had lacked sufficient time to prepare a counterargument to this request in time for today's hearing. Legal counsel for the claimant accuses CIC of making the process adversarial by manipulating the timing of document release. Bureaucrats and lawyers exercise power over space and time.

At the end of the hearing, the claimant's lawyer approaches me to explain why he was open to my presence as an observer. He suggests that the influence of the media on immigration policy has been so powerful that he has begun to encourage his clients from the boats to get more involved with the media, as he himself has done.

Refugee lawyers have repeatedly argued that the group arrivals detailed in this chapter made these migrants' access different compared with the great majority of claimants, most of whom did not arrive in Canada by boat. In response, CIC employees and IRB members argued that the unusual scale of these arrivals was, in fact, exceptional, and required unusual institutional arrangements to move claimants through the determination process in a timely manner.

What makes access to the refugee determination process exceptional? Does the quality of access vary, even though the system is intended to have each claimant's case heard individually? How do discursive identities manifest materially in the implementation of policy, access to the refugee claimant process, and the geography of detention? Through what processes are bodies and histories homogenized? This chapter pursues a strategy of embodiment by which geopolitics and the legal frameworks of asylum are "fleshed out" (Hyndman 2004; Mountz 2004). Corporeal geographies reveal the slippery nature of the boundaries of the state around the bodies of refugee claimants as they are constituted discursively through media representations and materially through the refugee determination process itself.

Given the findings on the importance of public image to the bureaucracy detailed in chapter 3, it would be difficult to grant too much weight to media representations when they are a key determinant not only of public opinion (Palmer 1998; Rivers and Associates 2000), but also of the orientation and operation of the federal bureaucracy. As bureaucrats struggled with the public perception that they were ineffective (Hier and Greenburg 2002), migrants struggled with the perception that they were bogus. Meanwhile, the dominant narrative taking shape through representations in the media was that of Canada's identity as a nation-state. Migrants posed a threat to a "leaky" nation-state. The process of identification that took place in the media, far from superfluous to the federal response to human smuggling, provided a narrative that explained the response and justified the geographies of access to the refugee determination process.

Eventually, the IRB accepted this woman's claim. This hearing itself was different from the vast majority of the hearings of claimants from the boats. Most were detained far away in Prince George and had their claims heard in provisional tribunals set up in the prison, as demarcated on Map 2. Almost all of those claims heard in Prince George were rejected. Among the 24 accepted from 577 claims made by those who arrived by boat (*Vancouver Sun* 2009), most were from women and children detained not in Prince George, but in the lower mainland region of Vancouver. The embodiment and gendering of the premigration experience recounted during the process disrupted the overarching story that this movement was composed of "bogus" refugees. The material evidence of this woman's case challenges prevailing assumptions that the boats carried primarily economic migrants. It also suggests that access may have been of uneven quality across dispersed institutional landscapes. The geography of detention may have influenced claimants' chances for positive findings by the IRB.

Asylum and Geopolitics

Asylum processes are always linked to current processes of securitization in international relations. During the 1990s, public discourse about refugees shifted from themes of humanitarianism, diplomacy, and the protection of human rights to security, criminality, and the abuse of "generous" asylum programs. These shifts were most shrill in the wealthiest Western nation-states with the highest rates of per capita immigration and refugee resettlement; these are also the nations perceived since September 2001 as the most vulnerable to terrorist attack. It is essential to study refugee determination processes in practice to understand how they are tied to processes of securitization.

Public opinion about immigration often coincides with displays of border enforcement (e.g., Nevins 2002). When public opinion turns against immigration and the perceived "abuse" of refugee programs, particularly in countries that feel vulnerable to terrorism, governments make a show of militarizing border enforcement. The performance of strengthening border enforcement that surges periodically along North American borders often compromises opportunities for asylum. This is but one example of the ways that asylum processes bring geopolitical relationships into view.

Didier Bigo (2002) makes a persuasive argument that politicians capitalize on the xenophobic fears of their constituents by basing political campaigns on the exclusion of immigrants. Often these narratives posit the precedence of national security over human security (see Hyndman 2004). Using the image of the "bogus refugee" is particularly powerful. The widespread perception of asylum claimants as fake and criminal enables ever stronger measures of exclusion in the form of interdiction, detention, and deportation.

This chapter illustrates how constructions of crisis by the state and the media give way to the making of the "bogus refugee," and how the narrative of the bogus refugee has powerful consequences for those seeking protection. This negative turn in public discourse accompanied the dramatic downturn in asylum seekers that has been observed since 2001.

In 1998 members of the Economic Development Corporation traveled to China to stage a dialogue regarding China's plans to join the World Trade Organization. After decades of isolation, China was opening up to trade with the West. Certain issues, however, had always hindered this partnership. Among them were human rights abuses such as persecution under the one-child policy and the suppression of dissident political viewpoints and religious practices. Experiencing these abuses is the most common basis of successful refugee claims made in North America by applicants from the People's Republic of China.

These abuses are what bureaucrats have referred to as diplomatic "irritants" over the years. Relatively high rates of acceptance among asylum seekers in North America embarrassed the People's Republic of China

(PRC), which has often accused nation-states such as Canada and the United States of encouraging economic migrants to apply for asylum by enacting overly generous refugee programs (*Vancouver Sun* 2000).

These issues served as a backdrop to the negotiations between Canada and the PRC over the refugee claims and repatriation of the Fujianese migrants. Chinese diplomats, for example, informed CIC that travel documents would be issued only for the repatriation of an entire boatload of migrants at once. This would serve as a forceful public message that these were not in fact Convention refugees fleeing persecution in the PRC, but rather economic migrants who had left China for opportunistic reasons and did not qualify for asylum in Canada.

Once CIC bureaucrats learned of this, some wanted to stop the communications branch from releasing the numbers of those on the boats to the public, to prevent China from knowing the precise number. The media were hungry, however, and the bureaucracy eager to feed. As a result, PRC authorities in Canada and China were well aware of how many migrants arrived on each boat.

As time passed, CIC managers and assistant deputy ministers learned how to navigate diplomatically to obtain travel documents. Eventually, on 10 May 2000, CIC chartered a flight to deport a group of ninety migrants back to the PRC, and it repeated this process with another group of ninety on 27 July 2000. These group removals overlapped with a visit of the Canadian trade mission to China, and affirmed complementary narratives in the media: that boatloads of migrants were deported because they were found not to be political refugees in Canada and therefore had not fled persecution in China.

There were glitches, however. Depending on their experiences with the refugee determination and appeals processes in Canada, it took individuals different lengths of time to make their way through the system. Under these conditions, it was a logistical impossibility for CIC to fill a plane with migrants from the same boat. Therefore, while the group deportations maintained the semblance of coming from the same boat, in actuality they were a mixture of people from different boats.

The story of the migrants' arrival by boat and their return by plane underscores the importance for nation-states to maintain a positive public image. Putting a group of migrants on a plane was an effort to homogenize rather than differentiate their experiences in order to convey a powerful, clean, unified message regarding human smuggling from China to Canada. It shows the pressure on, and the desire of, the bureaucracy to report a strong enforcement response to the public through the media.

Geopolitical relations influenced not only intergovernmental negotiations, but also the ways that the media represented issues, as well as—ultimately— the applicants' chances of having their refugee claims accepted. A successful asylum claim in North America reflects poorly on China by affirming that state-sanctioned persecution has in fact taken place there.

The episodes in which migrants were grouped on chartered planes for deportation illustrated the government's desire to homogenize this transnational migration, to impose an explanatory narrative that scripted meanings and identities onto migrants and states in relation to one another. The scripting of migrants as criminal projected Canada's identity as an enforcer and China's as an innocent bystander. One tale can be told from a distance: several hundred economic migrants attempted to cross borders illegally. Analysis of media coverage of the events reveals that the narrative of crisis gives way to the making of the bogus refugee. These "bogus" refugees were granted due access to Canada's refugee determination process but were ultimately deported. The system worked. And yet there are other stories to be told. A shift to the scale of the body to understand more about the state reveals interlocking processes of boundary construction around migrants, bureaucrats, and the nation-state, and the intimate relationship between body and body politic (cf. Campbell 1992; Nelson 1999). The individuation of experience uncovers power relations otherwise obscured and pokes holes in the dominant narratives deployed to explain the boat arrivals and other human smuggling movements by questioning some of their underlying assumptions.

Media Coverage and the Making of the Bogus Refugee

A cynical public discourse about asylum seekers has emerged, and with it the steady rise of the presentation of the asylum seeker as a "bogus refugee," which has had dire material consequences for all but the wealthiest people on the move to North America, Europe, and Australia. Asylum numbers have been dropping in these regions in recent years (Newland 2005), a trend that reinforces the notion that spaces of asylum are shrinking.

In Canada, when the boats were intercepted, the media inundated the public with images of a group that came to be known as "the boat migrants." Coverage of immigration stories becomes more pronounced during high-profile immigration events (Palmer 1998). CIC made front-page stories some eighty times that year (Charlton et al. 2002, 5). After conducting interviews with the reporters involved, Sorcha McGuinness cited this as one example of the ways that journalists were "swept along by waves of public opinion, dammed by institutional constraints and caught in an undertow of bias" (2001, 23). Recurring themes came to the foreground and resonated historically with well-worn tropes of Chinese immigration (e.g., Anderson 1991).

In addition to increased reporting, coverage of these arrivals signaled a qualitative shift in content, tone, and language, all of which contributed to what Hier and Greenberg have identified as a "discursive construction of a crisis" with "a capacity to recruit and mobilize newsreaders as *active participants*" (2002, 491) in the narrative of who composed this group of migrants, and who composed Canada in relation. Much of this crisis in the news unfolded as a "moral panic" in which the boat arrivals and the numbers of migrants were decontextualized, taking on a life of their own and hence becoming a "phenomenon" (Hier and Greenberg 2002, 503). Hier and Greenberg have argued that the state was implicated in this crisis in two ways: through the construction of the crisis ("weak" immigration and refugee policies) and through its resolution of it (a "strong" enforcement response) (2002, 492).

The media contributed to the crisis by doing little to contextualize the numbers as small (Clarkson 2000) compared with, for example, the number of people smuggled through Canada's airports, which was estimated in the tens of thousands annually (Interview, Ottawa, October 2001). The media portrayed the boat arrivals as "the last straw," with particularly notable headlines in the province, such as "ENOUGH ALREADY" (*Province* 1999c). Numerical decontextualization contributed to the sense of crisis and encouraged the notion that immigrants would potentially "flood" Canadian cities.[1] Despite the efforts of federal communications employees to present the appearance of being in control, the boat arrivals played out in the news as a crisis that provoked anxiety in the public. Media representations positioned the migrants as a threat to Canadian security, a fear exacerbated by images of crowding on boats.

The association of refugee claimants with criminality intensified as the story of 600 "illegal aliens" evolved; their motivation appeared to stem not from political persecution in their home country, but rather from the desire to pursue upward economic mobility (Greenberg 2000). The story was repeated so many times in newspapers that it seemed irrefutable: these migrants were "bogus," not genuine; they were "economic"—therefore opportunistic—and not "political." These constructions of criminality—also a common trope in media coverage of recent immigrant communities[2]—gained momentum throughout the migrants' time in Canada, from their portrayal through security fences at Work Point Barracks, to their movement in and out of correctional facilities and courts during the next eighteen months (McGuinness 2001, 20–21).

Refugee applicants were pictured frequently in handcuffs and prison uniforms (see Figure 7). While it is not in fact illegal to make a refugee claim, a visual and textual language of criminality ("illegal aliens") not previously associated with immigration to Canada accompanied media representations of the boat arrivals, with headlines such as "Detained aliens investigated" (*Vancouver Sun* 1999).[3] This language pitted a Canadian "us" against a foreign "them," as evidenced in the headline "Beware, illegal immigrants. We Canadians can be pretty ruthless" (*Province* 1999b).

The migrants came to be associated with the mode of transportation by which they arrived, an image indelibly recorded for the public in countless pictures of old, rusty ships carrying "boat migrants."[4] Early coverage implicated migrants' bodies as potential carriers of disease and other medical conditions such as malnutrition, dehydration, and hypothermia. Front-page photographs portrayed migrants crowded on boats with unsanitary conditions.[5] Leading articles such as one in *The Province*, with the headline "Quarantined," reinforced this image (*Province* 1999a). Newspapers placed language about public health concerns related to the boat arrivals on their front pages.

Whether through metaphors of disease, natural disaster, or criminality, the migrants were depicted as an invasion (Cresswell 1997) and a threat to national security (Bigo 2002). The Canadian government was shown as incapable of controlling its borders, overwhelmed by human migration, and guilty of having weak laws and a flawed bureaucracy (Greenberg 2000). Murky definitions of human smuggling suddenly seemed clear through processes of boundary construction and nation-building that linked the policing of borders to the policing of boundaries around migrant identities.

Illegality and the Rise of "the Bogus Refugee"

The arrival and eventual deportation of the Fujianese migrants, coupled with the announcement of new legislation in 2002—the Immigration and Refugee Protection Act—marked the culmination of a decade of tightened controls over immigration, during which time, remarked Sherene Razack, "The criminal attempting to cross our borders featured as a central figure in the discursive management of these new [federal] initiatives" (1999, 160).

The smuggled migrants were not the wealthy, "flexible citizens" (Ong 1999) whose migration Canada facilitates, nor were they represented as "genuine" political refugees. This increasing distinction between legitimate and illegitimate or criminal transnational migrants, conjuring up images of invasion, bolstered the apparent urgency of an enforcement response. The response of the government to human smuggling illuminated inconsistencies regarding the global positioning of Canada as both a humanitarian, refugee-receiving nation and a strong enforcer of borders to prevent uninvited migrants. The contradictory impulses to attract and repel represented the conflicting demands placed on the nation-state and the paradoxical postures assumed in response. This tension emerged not only in the media, but also in new immigration law designed to "open the front door and close the back door." It was in the day-to-day work of civil servants that the philosophical divide played out. It became evident in the uneven enactment of borders and policies, often tied to image.

In contrast with the discourses of globalization that involve flows of capital and elite business professionals across borders (e.g., Ley 2003), migrants, and especially poor migrants, are often characterized as leaks and invasions (Cresswell 1997, Cresswell 2006). This narrative parallels trends in Australia and the European Union, where it was also becoming increasingly difficult to be seen as a "legitimate" refugee. Increasingly in North America, Europe, and Australia, refugees and asylum seekers are pulled into the narrative about the loss of state sovereignty, and viewed as somehow "economic" and therefore not "political" or genuine.

Constructions of illegality in the media bolstered the discourse of the "bogus refugee." At the same time that states tightened controls on refugee movements, the media delegitimized those individuals who did succeed in making refugee claims. Bound up in this process was the artificial distinction between the economic and the political migrant.

Nation-states operate over large and diverse geographies to legislate citizenship by classifying and categorizing (Scott 1998), an exercise in power that, in the case of transnational migration, involves reading and scripting bodies as texts. Immigration policies prescribe varying degrees of national belonging. The operation of the federal bureaucracy corresponds with Foucault's theory of governmentality (1991) wherein governance of the individual moves into the realm of the social body, where deviant identities are produced and policed. While fulfilling the mandate to protect, CIC proceeded in a context in which this group of claimants was presented as distinct from others in terms of their mode of travel. The media contributed substantially to this climate. Millions of dollars were mobilized to the response, and the state entered the business of detaining refugee applicants.[6] This was the environment in which the new immigration and refugee legislation was proposed in the House of Commons only a few months later. It was also the environment in which the IRB adjudicated the claims made by the 1999 arrivals, such as the one with which this chapter opened.

In the context of human smuggling, Canadian identity and nationhood were constructed in opposition to, rather than through, the promise of immigration. Regarding changes to Canadian immigration legislation in the 1990s, Razack (1999, 160) argued, "One of the paramount tasks of border control, and the justification for all new initiatives, became the separation of the legitimate asylum seeker or immigrant from those deemed to be illegitimate." This was the narrative formulated in the media. Several months after the boat arrivals, these contrasting narratives of "good" and "bad" immigrants crystallized in the language of new legislation to preserve the promise of Canada's future through immigration, facilitating the entry of desirable immigrants while protecting the nation against illicit movements.

The discourse surrounding the boat arrivals from China reflected these efforts to categorize according to legitimacy. The magnitude and tenor of the public response categorized this group of Fujianese migrants as particularly egregious. Eventually the representations settled into repetitive construction of the binaries—"deserving" and "undeserving" immigrants (Razack 1999). In contrast with business immigrants, welcomed for the promise of their economic contributions (e.g., Mitchell 1993; Ley 2003), the "boat migrants" were construed as greedy. Moreover, implicit in the processing of this group was the representation of them as disorderly.[7]

Nations manage populations by producing identities discursively through practices of classification and categorization (cf. Scott 1998), exercises that entail the material inscription of identities onto the body (see Pratt et al. 1998, 1999). The media contributed to the regulation and surveillance of migrant bodies in relation to popular interpretations of immigration policy. The narrative of the illicit entrant affirmed the story of the violation of what was perceived as a nation too generous with its immigration policies (Razack 1999, 173). This contributed to the environment in which migrants experienced the refugee determination process. It is, therefore, important to think about the ways in which governmental policies and procedures unfold: never in a vacuum, but rather amid social and cultural contexts. The refugee determination process unfolded within contexts of criminalization, as the media portrayed handcuffed migrants in detention and in BC prisons.

Access to the Refugee Determination Process

The "long tunnel thesis" detailed in the Introduction represents the relationship between identity and geography. The lengthy debate about when the claimants would have access to legal counsel began during processing at Esquimalt. Lawyers might advise migrants to make a refugee claim and to relay their history in a certain way, whereas CIC wanted unfettered access to the migrants without counsel in order to better understand their individual cases and the details of their journey. With the long tunnel, the government reworked the local geography by temporarily designating the base a port of entry (POE). Thus they declared the migrants in the stage of being processed, as though they were walking through the long tunnels of an airport, rather than as already landed on Canadian soil. These representations demonstrate struggles over language, geographies of access, and legalized geographies of power (Blomley 1994) that continued throughout the refugee determination process and emerged repeatedly in interviews with lawyers, who described the process as exceptional.

The first major contention of the lawyers in BC was the way the definition of detention was subverted by CIC's designation of the Work Point Barracks at Esquimalt as a POE. Migrants in detention have a right to legal counsel, but migrants being processed at a POE do not have such a right, based on a Supreme Court of Canada ruling. In interviews with me, and also publicly during that time, lawyers asked under what circumstances a migrant was considered detained. For them, the presence of barbed wire, dogs, and RCMP officers on a military base signified detention. If detained, a person had a legal right to counsel (Interview, Vancouver, August 2001). What had the hallmarks of detention—guard dogs, barbed wire, and commissionaires—was officially designated a POE instead. The migrants were processed for several days at Esquimalt. The information gathered in preliminary interviews was later used by the government in its opposition to the asylum applications, on the basis of contradictions in stories. These were the interview notes that the refugee lawyer argued should be excluded from the hearing in the opening of this chapter. He succeeded in excluding the notes from his case.

CIC, in the meantime, was interested both in providing due process and in learning how to do this most efficiently. The department was also trying to learn as much as possible, as quickly as possible, about how the migrants had been smuggled to Canada. Refugee lawyers agreed that the migrants should have showers and medical exams, but argued that there was no need for this process to be prolonged up to fourteen days, the amount of time that it took for some migrants to be processed at Esquimalt (Interview, Vancouver, August 2001).

Normally, lawyers who represent refugee claimants are retained by clients through Legal Aid, to which they are generally referred by immigration officers at POEs. In this case, however, lawyers became involved in a nonconventional way. Rather than being retained by refugee claimants through Legal Aid, some lawyers chose to involve themselves because of their belief that these migrants should have access to legal counsel early on. Perceptions of social injustice motivated them to become involved regionally. They organized collectively through the Refugee Committee and the Immigration Section of the BC branch of the Canadian Bar Association.

> The Immigration Bar is deeply concerned with the position of the Department of Citizenship and Immigration Canada that the migrants have no right to counsel for the purposes of their initial interview and Senior Immigration Officer (SIO) interview. The Bar is concerned that migrants do not understand the law, and do not understand that they must initiate their claims before the SIO interviews conclude. (*BarTalk* 1999, 1)

This debate about the beginning and end to processing and detention pertained to time, space, and access. Lawyers asked whether migrants had understood the opportunity and method by which to make a refugee claim.[8] They suggested that access to human rights was being subverted by the manipulation of space. CIC employees, on the other hand, argued that they were moving quickly in order to protect access to the determination process without keeping claimants in detention for unnecessarily extended periods of times.

After processing at Esquimalt, migrants from the second, third, and fourth boats were detained in a provincial prison located in Prince George, in the interior of the province, a ten-hour drive from Vancouver where most services for immigrants and refugees are based. Lawyers argued that the aberrant conditions characterizing the refugee determination process related to the identification of the migrants as illegitimate, illegal, and bogus, and also to the unusual location of their detention. Lawyers also argued that claimants moved through the system as a homogeneous group. At routine detention review hearings, federal lawyers used the same document to provide background information on each claimant and to argue for continued detention, which lawyers and advocates suggested was a form of racial profiling.

As the refugee determination process unfolded institutionally, the process of criminalization ensued in the domain of the public court—the mainstream media. Like CIC, advocates and refugee claimants themselves battled the weight of media representations as they saw their refugee claims through the process in Canada. In interviews, lawyers and refugee advocates argued persuasively that these refugee claimants experienced a refugee determination process that was not routine, but was aberrant in terms of the geography of the process, access to legal representation, and their identification and treatment as a group rather than as individuals.

One lawyer for the claimants described the weight of media representation:

> The media played a massive role in criminalizing these people by referring to them as illegal migrants. Also, the fact that they were seen to be in detention added to the specter of illegality. We tried to get out the fact, and I think not always very successfully, that refugee claimants are not illegal. They are not operating outside the law.
>
> According to our laws, a person has the right to make a refugee claim after arriving in Canada—regardless of how they get here. To come to Canada and make a refugee claim is not illegal. That is not an illegal refugee claimant. It was something that I don't think was ever properly explained in the media or elsewhere. The illegal aspect is that they arrive without valid documents; they arrive without a visa, but they make their refugee claims after arriving in Canada in accordance with Canadian law. They're not jumping queues, but following a set process for persons who claim to have a fear of persecution. (Charlton et al. 2002, 15–16)

These statements affirm that legality and criminality shift meaning across time and space. Geographic location corresponded with identity: "bogus" refugees were held at bay far away, where it was more difficult for them to access resources based in Vancouver.

In their struggle against media representations, lawyers linked the response of the federal government to the process of identification that took place in the public domain, thus enhancing the connections between boundaries around the nation-state, governance, and migrant identities:

> The overarching frame in all of this was, "Oh, we know they're economic migrants, and we have to give them due process. We will give the appearance of due process, fund due process. But what does that really mean?" (Interview, Vancouver, September 2001)

This lawyer distinguished further between "thick justice" and "thin justice" and argued that "the substantive quality of justice is what I think justice is about." Due process gives claimants representation, the right to speak, and physical access to the system. But, he argued, those claimants detained remotely in Prince George experienced a form of "skeletal justice" rather than the "thick justice" more readily available in Vancouver. He accused the government of "profiling by pushing people through the system . . . but the frame doesn't hold" (Interview, Vancouver, September 2001).

Statements like these suggest that the systems and policies of the nation-state are implemented unevenly across space and time, that different groups have differing levels of access to the nation-state, and that a geographical analysis of "the translocality of state institutions" (Gupta 1995, 392) uncovers the processes by which the state shapes access to protection. Nation-building practices inherent in the regulation of immigration policies are not only bound up discursively with identity construction, but also experienced materially by those denied entry.

Lawyers and bureaucrats raised important questions in interviews about the relationship between media coverage and the quality of access to the refugee determination process. In interviews, bureaucrats described harassment in the broader community for what was perceived by the Canadian public as generous treatment. They told stories of being challenged in coffee shops and dentists' offices alike for not simply sending people home without due process. To what extent did this continuous media pressure, on both an individual and a collective basis, influence the process?

In BC, the political will for detention gave way to an expedited enactment of the refugee determination process. One lawyer remarked:

> The system was very eager to contain, and the system was eager to process them on an expedited basis. And I would say with a desired outcome. The frame was that these were not actual refugees. The frame was that these were economic migrants. . . . The question then becomes, how do you contain six hundred people. . . . Build a frame; work within that. Anything that leaks out, push it back in. (Interview, Vancouver, September 2001)

The intense shaping of migrant identities for the public related powerfully to their access to due process. Refugee advocates and lawyers criticized CIC for the stress of long-term detention, for the criminalization of refugee applicants, and for the geography of detention.

Locations where claimants are detained influence the quality of access, and the differing experiences of individual claimants going through the process uncover problems with the system as it operates across distance and time. The claimant whose hearing was a stage for legal debate was excluded from making a refugee claim in Canada during her initial interview at Esquimalt. Once she was included, the government used notes against her case that had been gathered during her preliminary interview, before she was granted access to a lawyer. In spite of the consistent attempts of the federal government to support the rejection of the claimants from Fujian, she was one of the small group ultimately granted refugee status in Canada.

Detached Geographies of Detention

The lawyers' litany of complaints in representing this group highlight legal geographies of power (Blomley 1994) and prompt an important question as the detention of asylum seekers is on the rise globally. To what degree do institutional geographies—of access, detention, advocacy, and service provision—mediate the quality of access to rights?

The physical placement of detention centers is a material expression of the idea of the "bogus refugee." While the federal and provincial governments argued that logistical planning and financial considerations determined the location of detention facilities, advocates and lawyers for the migrants clearly did not accept this argument and accused them of choosing areas of convenience that best served their own objective—to *not* grant refugee status.

Most refugee claimant hearings in British Columbia take place in Vancouver. Over time, most migrants from the boats were detained at the Prince George Regional Correctional Centre in the small interior city of Prince George, an especially difficult drive from Vancouver in the winter. One lawyer based in Vancouver explained it this way:

> And once the issue of detention came about, it was a question of where they were going to be detained en masse. They were detained outside of the Lower Mainland[9] in Prince George. This was very significant because again it limited access to legal counsel for the detainees. Prince George is halfway up the province. There aren't a lot of refugee lawyers practicing in Prince George. There are also not a lot of certified interpreters and translators in Prince George. (Charlton et al. 2002, 14)

Once determined, this institutional geography influenced subsequent decisions. Far away from refugee lawyers, interpreters, resettlement agencies, human rights monitors, and the regular tribunals of the IRB, special accommodations had to be made for processing. Lawyers called these contingency plans "disastrous" (Interview, Vancouver, September 2001). Hearings normally take place in the chambers of the IRB in Vancouver; but for this group, they were held in provisional tribunals established within the prison in Prince George and adjudicated by officers of the IRB who were flown in. Claimants attended hearings in prison uniforms and handcuffs. Separate detention facilities in Esquimalt and Prince George limited lawyers' access to clients and the claimants' access to due process. So the way the state perceives human smuggling has powerful material, geographic, and legal ramifications for displaced persons.

Legal Aid, overwhelmed by the number of claimants in need of representation and not accustomed to servicing this number of clients simultaneously, initiated a process wherein lawyers bid for contracts to represent large numbers of claimants. The Legal Services Society (LSS), which funds Legal Aid, reviewed bids by lawyers to represent a block of clients for a variety of fees.[10] While LSS never released the rationale for which four contracts were subsequently awarded, experienced refugee lawyers I interviewed noted that the four lawyers chosen were not ones who regularly attended the immigration subcommittee meetings of the Canadian Bar Association, nor were they generally well known to the community of refugee lawyers in Vancouver (Interviews, Vancouver, August and September 2001). A series of controversies surrounded the competency and experience of these lawyers, as well as their ability to represent at once such a large caseload of clients to whom they had so little access because of the location of detention.[11]

Without proof, the legal community speculated that the bids had been awarded for cost (Interview, Vancouver, August 2001). The more experienced refugee lawyers who were *not* selected questioned the criteria by which the four *had* been selected, the quantity of clients represented per lawyer, and correspondingly the quality of representation for those imprisoned in Prince George. According to one lawyer, the controversy "bears out the worst fear. I think that Legal Aid failed" (Interview, Vancouver, September 2001). To him, this represented an institutionalized failure of the system for this set of claimants.[12]

Lawyers also condemned the detention review processes in Prince George. Refugee claimants in detention have routine detention reviews, during which time federal lawyers argue the case for continued detention, and the IRB adjudicates. As routine detention reviews took place for claimants, lawyers argued that they were treated as a homogeneous group. CIC legal counsel routinely presented the same document at the reviews that profiled migrants from Fujian, which one lawyer argued was evidence of racial profiling.

> It was very clear with this group of people that they wanted to contain them. So it looked to us like they were practicing racial profiling. But of course they took great exception to that. . . . I asked them, "How would you describe this book? You're profiling this group. You're detaining them as a group; you're processing them as a group."

He described the book as a series of color photocopies of rusted ships; once again the group was identified according to mode of transit. He suggested that the minister's representatives argued for continued detention with a "standard script" and a statement about why Fujianese migrants traveled to Gold Mountain (Interview, Vancouver, September 2001).[13] He argued that the claimant process was supposed to "individuate" but that, in actuality, names and dates of birth were the only distinctions made among clients' files in detention reviews for this group.

With no offices in Prince George, members of the IRB were flown in and temporarily accommodated. Advocates also asserted that the IRB could have better serviced the refugee claimants by "standing up" to CIC and insisting that the hearings take place as usual in Vancouver, rather than forcing them to move offices and staff to Prince George.[14] They argued that the IRB adjudicators disliked this experience, wanted to assess cases quickly and return home, and were therefore biased by the geography of the process.[15]

Lawyers argued that the cases moved in and out of the system at a faster rate than normal.[16] Because the hearings were expedited, legal counsel argued that they did not have as much time as usual to gather evidence for the cases. From the IRB's perspective, however, the cases were prioritized *because* the claimants were in detention and, in the case of minors, because they were unaccompanied. The IRB found itself in a catch-22 scenario, challenged to oversee cases fairly and in an expedited fashion because of advocates' criticism of the federal government for detaining refugee claimants for long periods of time (Interview, Vancouver, August 2001).

Some attributed this to the fact that, like other institutions, CIC was overwhelmed and attempting to "contain" the situation:

> It was obvious that CIC did not have the capacity to deal with this group of people when they arrived. . . . What I noticed was a series of system failures that ensued. The question then becomes, how do you contain 600 people? Well, you contain with an army base, guard them with guns, search them. They have to target them, contain them, investigate them. (Interview, Vancouver, September 2001)

This theme of containment continued with restricted access over time. "Immigration as a part of its containment strategy decided to move people to Prince George" (Interview, Vancouver, September 2001). In Prince George, they noted that migrants were isolated, removed from the Chinese community and advocates in Vancouver.

What was the outcome of these detached geographies of detention? The location of the claimant process gave way to an interesting outcome. In 1999 China was the second-largest source country for positive refugee claims in Canada, with a 58 percent approval rate (United States Committee for Refugees 2000a). The rate for those who arrived by boat in 1999 was less than 5 percent, with only twenty-four positive claims granted (United States Committee for Refugees 2001), and several of these positive claims were challenged by the minister.[17] Most of the claimants who had been detained were ultimately repatriated in 2000, once they had exhausted their options for appeals through the courts.

Two women among the claimants from the boat arrivals were depicted in the *Globe and Mail* in 2000, standing on the roof of their lawyer's office

Figure 8. Two claimants from the boats intercepted from China stand on the roof of their lawyer's building in downtown Vancouver. Their identities are concealed, faces down, as they hold pins of the Canadian flag. Published in *Globe & Mail* on 22 July 2000. (Courtesy of photographer Lyle Stafford.)

building in downtown Vancouver (Figure 8). One of the women had her claim denied, and the other was accepted. The ninety adult females on the boats composed only 15 percent of the group, but received more than 50 percent of the initial positive claims. There are two possible explanations. The first has to do with gender. As noted at the beginning of this chapter, some argued successfully that they had faced persecution under China's one-child policy. Others received successful claims on the basis of persecution for religious practices.

The second argument has to do with geography. The majority of the twenty-four positive claims were granted to women and minors, housed in prisons and youth facilities in the greater Vancouver area. Here they were able to access better services, interpreters, advocates, refugee lawyers representing only a handful of claimants each, and the standard, downtown tribunals of the Immigration and Refugee Board rather than the temporary tribunals in the prisons. According to legal representation for the claimants, the IRB heard these cases in a more individualized fashion. While the process appeared to "work" for detainees in Vancouver, it appeared to fail those detained in Prince George. The question remains whether, with more equal access, there would have been more than twenty-four successful claims.[18]

As with the group deportations by plane, a more disembodied narrative tells a simpler story: the arrival was of a large group of economic migrants who were not political refugees, an assertion supported by the outcome of the claimant hearings. But a closer look at the embodied and located experiences of claimants moving through the refugee determination process shows that the story is more complex: those moving through the process in different locations had unequal access to the system.

Civil servants complained frequently of being outdone by human smugglers who could cross international borders with relative ease. Bureaucrats inhabited more bounded spaces, finding themselves confined by bureaucracy, boxed in by political imperatives, suppressed by the media (which they characterized as "out for blood"), and—more than anything else—restricted by the law. They felt oddly powerless.

The powerlessness expressed by bureaucrats and noted in chapter 3 extended to their engagement with the law. According to Conquergood (2002, 145),

[D]e Certeau's aphorism, "what the map cuts up, the story cuts across," also points to transgressive travel between two different domains of knowledge:

one official, objective and abstract—"the map"; the other one practical, embodied and popular—"the story."

There are all sorts of official and unofficial stories told about international migration, about refugee claimants, about human smuggling. There are many different domains of knowledge. There are maps and stories, stories and actions. It is essential to examine both domains of knowledge—in the form of policy and practice—in order to understand the plight of those seeking asylum.

Of course, civil servants, though they felt powerless, were in fact powerful. Like smugglers, they enacted informal transnational networks that smoothly crossed the boundaries they considered so constricting in their daily work lives. As ethnographers, we must pay closer attention to the discrepancies between what civil servants say and what they do (Herbert 2000; Mountz 2007), between how the state appears and how it is actually enacted in daily life, between the texts of policies and their actual implementation. In this case, we must note the differences between sovereign territories delineated on the map and the alternative geographies where the state acts and detains. These discrepancies are key to understanding the contemporary state: the state that has been devolved or "hollowed," the state in crisis, the state that performs border enforcement, the state that acts transnationally.

Like facts, terms, and policies, laws are social constructions of their era—written, interpreted, and reinterpreted over time as contexts change. Speaking historically, it is only in recent years that nation-states have assumed the right to determine laws (Heyman and Smart 1999) and to impose them on human mobility. In Canada, as elsewhere, immigration laws are known to be designed somewhat loosely, but with heavy regulation through implementation that allows the state a modicum of flexibility over time.

One CIC authority made this comment: in their work, laws tend to get in the way, and they have to try to accomplish what they need to despite the laws (Interview, Vancouver, June 2000). While the department maintained its commitment to upholding the law, at the same time it seemed to manipulate the law through its strategic legal interpretations.

Discourse, Geography, Identity

States are seizing moments of "crisis" to justify tactics to keep displaced persons elsewhere, away from sovereign territory. This seemingly localized geography in British Columbia is connected with trends occurring internationally, which are detailed in chapter 5. Long before a refugee claimant attends her hearing in the tribunals of the IRB, her case is tried in the court of public opinion, where she has little to no representation. The configuration of identities through public opinion and the media manifests in struggles over access to sovereign territory and refugee programs. The story of *who* the migrants were justified their location. Unfortunately for those seeking protection, refugees are being melded with terrorists, so that all forms of human migration become suspect and securitized. The very notion of a genuine asylum seeker in need of protection is disappearing from the public lexicon as states enact strategies to contend with the bogus refugee.

Boundaries are policed around the definition of the "Convention refugee." Those designated outside the definition are detained where their access to sovereign territory (see chapter 5) and claimant processes is inhibited, illustrating the relationships between discourse, representation, and identity. Discourse is central to the transformation in political power from sovereign to disciplinary practices (Foucault 1991, 1995). As the migrants arrived, parties such as CIC, refugee advocates, immigration lawyers, and the media each played a part in their identification. The lexicon of immigration perpetually evolves to reflect changing political contexts in receiving nations (Ellis and Wright 1998; Nevins 2002). In the late 1990s refugees and asylum seekers in North America, Western Europe, and Australia increasingly received the nomenclature of "illegal," "bogus," and "economic migrants" and were therefore often seen as illegitimate refugee claimants. Following the boat arrivals in Canada, the press brought "illegal alien"—a term more commonly used to refer to unauthorized migration to the United States—into circulation (McGuinness 2001). "Transnational organized crime" also became a common buzzword. A trend toward the criminalization of refugee applicants has accompanied the shifting channels of international migration flows, which are increasingly associated with human smuggling.

Increasingly, a person's location is bound up with his or her criminalized identity, and nation-states are making it more difficult for transnational migrants to reach sovereign territory to make asylum claims. Asylum seekers migrate through ever-thickening, geopoliticized fields of securitization. The shift in discourse corresponds with changing geographies of border enforcement practices and the resulting reduction in the number of asylum seekers in Europe, North America, and Australia.

Between Body and State

Chapter 3 embodied the state quite literally, to illuminate who enacts policy and in what way. This embodiment entails institutional research that de-stabilizes the notion of the nation-state as a monolithic entity and offers a deeper understanding of bureaucracy—a key institutional apparatus of the state, rarely dissected in research (Hansen and Stepputat 2001, 17). Embodiment of the nation-state thus follows a feminist strategy (Harding 1986; Haraway 1988; Butler 1990): to locate power relations and to contextualize decisionmaking with a workplace, a life history, and a social geography.

A second approach to the body is the Foucauldian notion that power produces identities through discourse, that identities are inscribed onto the bodies of migrants and bureaucrats (see Pratt in collaboration with the Philippine Women's Centre 1998; Pratt 1999).[19] These discursive practices of identification, categorization, and nomination are exercises in power. The state views migration (cf. Scott 1998) through the lens of its ability to assign and classify, functions bound up with governance and governmentality. Tracking the interdiction and detention practices of refugee-receiving governments demonstrates both visually and textually that the geography of these governance practices is shifting and made possible through the assignment of identity in public discourse.

The inscription of identity is most visible in the mainstream media, where oversimplified renditions of complex events circulate as explanatory narratives. Like civil servants, media workers are powerful players in the politics of identification and representation of transnational migrations.

Human smuggling and trafficking serve as particularly poignant affronts to the nation-state because they involve an explicit challenge to border enforcement. These industries involve the movement of people as commod-ities for consumption in the global sex trade and other service economies for a substantial profit.

Images of immigration often narrate the story of the emasculated nation-state circulated by civil servants—one that is rendered powerless by out-of-control immigration, embodied by migrants who materialize in dis-course through metaphors of invasion, flood, and waves (Ellis and Wright 1998). This discourse of leakiness with regard to North American borders was on the rise for some time, and intensified after 9/11. Consistent with Saskia Sassen's argument that, in an age of globalization, nation-states flex their muscles at borders (1996), the Fujianese migrants were represented in the media as a challenge to Canadian sovereignty and its ability to police its borders. This emasculated state acting in defense of its territory undertakes offshore activities that inhibit migrants and potential refugee claimants from reaching sovereign territory.

Even in Canada[20]

Whereas Canada was once known as a more progressive, humanitarian state in its granting of refugee status, it was now being portrayed in the media as soft, with the integrity of its refugee program threatened. The cartoon in Figure 9 depicts the Canadian government as a marine filling station, offering welfare assistance, a lax court system, and the acceptance of hard-luck stories.

This image illustrates a notable shift in Canadian public opinion toward immigration, public discourse surrounding immigration, and the political will and ability of the government to respond. The media produced these images for consumption by an anxious Canadian public; the migrants were symbolic of the perceived laxity of border control. Not only were they seen

Figure 9. This cartoon represents Canada as a marine filling station that offers welfare, a lax court system, and express service to "bogus refugees" such as those crowded on the approaching ship. (Copyright Adrian Raeside; originally published in the *Victoria Times Colonist*.)

as crossing borders in unsavory and disorderly ways, but they were perceived as a burden on the ever shrinking welfare state.

The border around the nation-state serves as a symbolic expression of the well-being of the Canadian social body (see Gilbert 2007). Migrant subjectivities are bound up powerfully in narratives of nation building and immigration policy, positioned as the foreign body that compromises the home. Cresswell (1997) presents an image of society as a human body, and leaks as out of place, in need of being cleaned or removed. This describes well the containment practices with which the federal government reacted to the Fujianese migrants. In response to national and international public pressure to patch its "leaky" borders, and to counter media representations of an at-risk social body, the government worked to put on a show of control (see Cresswell 1997; Hage 1998; Nelson 1999).[21]

With the nation-state represented as a leaky body, the construction of boundaries around identity is one strategy for containment of these leaks. Such postures have powerful material ramifications for refugee claimants and other displaced people, notably in terms of the decision to detain. As lawyers suggested, however, these practices of containment extended beyond detention of the migrants themselves. The federal government contained the messages to the media, through which it retained the power to identify and classify (Scott 1998).

Simultaneously, practices of containment were unfolding within the government itself, where a strong effort was being exerted to control the public image of the federal response. So the nation-state exercised control around identity construction to depict neat narratives about human smuggling and effective government responses. But processes of abjection (Kristeva 1982) are always less concrete and more ambiguous than government's attempts to categorize. There is more gradation (Anzaldúa 1987; Salter 2004) along the border than the lines on a map or the law would have us believe. The containment strategies were an attempt to delineate between black and white, between the inside and the outside of the nation-state. But such clean dualities between self and other never hold. Scott's (1998) view of how the state sees and manages works only from a distance, not up close. Processes of identity construction on the ground are far messier. Binaries bleed. The state is here, there, and everywhere.

Power, too, bleeds through these embodiments of the nation-state, and it is therefore important to analyze who is embodied, how, and why in the relationship between the state and smuggled migrants. These embodiment schemes reveal the processes through which the power of the state materializes in everyday life. To some extent, the embodiment/ disembodiment of the state is a function of a large bureaucracy that suppresses and normalizes the individual by emphasizing the whole. Bureaucracies are designed to manage information as well as public image, and this usually takes the form of written policy, organizational diagrams, and press conferences. In this sense, the government often appears to be transparent on paper and in the media.

Although immigration bureaucrats usually work behind the scenes, the boat arrivals drew officers out of their offices and onto the water in ways that made them very visible to the Canadian public. Federal officials were portrayed in particular ways. Covered in white biohazard uniforms with hoods, black vests, and black boots, they were standardized and secure, an embodied expression of the state in control of its borders. The state is therefore strategically and visibly embodied in distinct ways in relation to different policies and populations.

Moira Gatens challenges monologic narratives of the state by questioning the utility of the "body politic" as a metaphor (1991). "The metaphor functions to restrict our political vocabulary to one voice only: a voice that can speak of only one body, one reason, one ethic" (81). She notes that if the body politic is ascribed only one voice, "any deviation takes the form of gibberish" (85).[22] Nigel Thrift suggests, "The lack of embodiment, except as inscribed representation, perhaps explains why critical geopolitics has produced so little work of an ethnographic bent" (2000, 383).

An *embodied* state appears less powerful, more vulnerable, and a bit unprepared to respond to smuggling. Locations of power show attempts of the nation-state to strategically mediate transnational globalization, mobility, and displacement. Embodiment of the nation-state moves beyond policy analysis to locate political processes in the more fluid, daily, personal interactions that surround and disrupt these formal instruments of governance.

Embodiment is a strategic political alternative to the ways in which popular discourse, the mainstream media, and governments push the migrant body to the foreground while concealing the embodiment of the state.[23] The work of locating the nation-state through corporeal geographies disrupts the understanding of the state as a comprehensive, coherent project. On the contrary, state practices materialize distinctly at distinct moments for distinct groups of people in differing political contexts. Bodies are one locale where the shifting geographies of governance and the reconstitution of mobile borders are brought into relief. As opposed to the existence of a coherent, hidden strategy of the state awaiting discovery, there is a group of diverse, conflicted bureaucrats themselves often trying to figure out the objectives of state projects. Analyzing state practices at multiple levels opens the potential for new understandings of the roles of nation-states in global processes.

Containment

States try to contain bodies, spin stories, and control information. Detention proved, in part, an attempt to sustain the integrity of the government in the minds of the public and to foreign governments. In interviews at CIC, respondents mentioned a clear demand from the public to "do something." Detention is among the most expensive, reactive, and short-term solutions to undocumented migration, but it is a visible expression of a swift government response, of *containment* of the problem. Images in the newspapers of a government not in control of its borders soon gave way to images, such as Figure 7, portraying minors from the boats in handcuffs, moving in and out of detention centers, of bodies contained, of a situation brought under control. In his analysis of Australian public discourse about immigration, Ghassan Hage referred to detention as "ethnic caging" (1998, 105), the ultimate physical expression of racialized "othering."

This strategy was an effort to reach the unidentified human smugglers. By detaining the migrants in order to impede the smugglers, the nation-state imprisoned one set of people to deter another. While smugglers remained unidentified and hidden from view, their clients were overidentified, held captive as a visible and costly message to smugglers and the outside world. Detention communicated to several audiences—including potential migrants in China, the Chinese and American governments, and the Canadian public—that human smugglers would not operate successfully in Canada, that Canada would respond with a show of force and maintain its ability to police its borders. In their absence, the state painted a portrait of human smugglers in a narrative that involved a performative response. Migrants, of course, also figured prominently in this narrative, embodied in particular ways.

In media coverage of the boat arrivals, the narrative of the illicit entrant affirmed the violation of what was perceived as a nation already too generous with its immigration policies. These perceptions manifested in the language and images of the movement, and contributed to the environment in which claimants experienced the refugee determination process. The problem with a government driven by media coverage is that it leaves no room for reflective dialogue, the admission of mistakes, or exploration of alternatives. Instead, nations respond quickly and capitalize on moments of crisis as communicated by the media.

Relationships among nations and borders materialize bodily most obviously for the displaced person or migrant. But an embodied state reveals other sites of global struggle suppressed in our narratives of transnational migration. Through everyday, local geographical analysis, embodied experiences show struggle on the part of migrants, bureaucrats, and others, despite the powerful homogenizing effects of media representations. Much of the success of the Canadian response was measured in terms of the effective communication of images. Bodies were caught up in these excessive efforts to project certain images. Bodies mattered so much precisely because of their visibility—and invisibility—in different times and places, and their resilient presence in spite of these efforts. Analysis of the state through the body uncovers a powerful fixation with inscribing identity onto the migrant deemed political or economic.

The strategy of embodying the state draws on feminist theories that locate knowledge in a time and a place. This location is accomplished by conducting analysis at multiple levels—in this case, looking at transnational flows from the perspective of the nation-state, and at the nation-state from the perspective of the body. Research of the embodiment of the state uncovers the inconsistencies of dominant discourses of migration. Examined through a finer lens, the state is not quite able to contain identities through boundary making. As an alternative to Scott's depiction of the ways in which the state sees and categorizes human mobility from a distance, we can analyze the nation-state through the body. We can dissolve boundaries by locating their reification around the body of individual migrants and immigration bureaucrats alike, in their day-to-day experiences where they negotiate state constructions of identity.

The state is being reorganized geographically and discursively through the production of migrant identities. The management and scripting of bodies proves central to the body politic (Nelson 2004) in building corporeal geographies of the nation-state. It is in the blurring of boundaries at sites of nation building, governance, and identity construction that there is potential for political change.

This chapter has traced the relationship between productions of identity and geography—here, the discourse of the bogus refugee related to the remote detention and subsequent invisibility of those seeking asylum. It showed how states capitalize on moments of crisis as communicated by the media, and how global processes of exclusion are inscribed on migrant bodies. The narrative of *who* these migrants were explains *where* they were located and vice versa. Boundaries function in international political territorialization and in practices of identity construction.

It is essential to understand that this anatomy of crisis construction was not an aberration (as it appeared to be in Canada at the time in many ways), but part of a larger, international trend. Chapter 5 documents this trend and offers examples from other countries and regions with significant refugee resettlement programs: Australia, the United States, and Europe.

Chapter 5   Stateless by Geographical Design

CHAPTER 3 EXAMINED the bureaucracy, one node in a transnational network where civil servants manage migration. Chapter 4 explored the intricacies of access to the refugee determination process once people had landed on sovereign territory. This chapter moves farther out still from the center to dwell in offshore zones that render migrants stateless by geographical design. I use the term "stateless by geographical design" to signify extra-territorial locations that are neither entirely inside nor outside of sovereign territory, but that subject migrants to graduated degrees of statelessness by introducing ambiguity into their legal status.[1] This restricts, to various degrees, their access to national and international human rights laws. Hence, these zones are also legally ambiguous, as exemplified by the struggle over the denial of access to refugee lawyers in Canada. Such sites are created by states to limit migrant access to sovereign spaces and systems.

Giorgio Agamben (1998) traces the history of the Western camp as a zone of exception, where inside and outside realms of sovereignty are blurred. These "exceptions" are central to the construction of a time of crisis, during which—in the case of border enforcement—the contradictions of border policing come to the fore (e.g., Andreas 2000; Nevins 2002).

Agamben's work also suggests the recategorization of migrants in relation to states; in particular, those excluded from the state during crises are what Agamben calls *homo sacer,* "bare life," persons that the state has stripped of all rights. Here I link the discursive construction of the "bogus refugee" with its corresponding geography: the respatialization of asylum, the dwindling numbers of asylum seekers in Western countries, the growing numbers in detention, and the increased enforcement offshore.[2] The gradual decline in asylum claims corresponds with trends in policies involving detention, border enforcement, and criminalization of displaced populations. As governments capitalize on security concerns, refugees fall by the wayside, quite literally, detained in marginal zones of sovereign territory. Whereas chapter 4 exposed the anatomy of exclusion in times of crisis, this chapter shows that zones construed as exceptional in the national context are not so when contextualized internationally. Measures taken in the name of "homeland security" are fast approaching a critical moment of crisis for refugees globally. This may well lead to reconfiguration of the very laws and international agreements through which people seek protection (See Macklin 2005).

For asylum seekers and undocumented migrants, the state looms ever larger since the terrorist attacks in 2001. Roger Waldinger and David Fitzgerald (2004) argue, "In the long view, the rise of massive state apparatuses *controlling* population movements between states represents the most striking development" (1188). With scholars now in agreement that the regulatory power states exercise to control migration has been overlooked, their next critical step is to study *how* states regulate migration.

Throughout the 1990s nation-states pursued more aggressive enforcement strategies, and migrants increasingly employed the services of human smugglers (Kyle and Koslowski 2001). As the industries of detention, enforcement, and human smuggling grew, both states and smugglers exploited fluid spaces between countries, utilizing creative geographical strategies to outmaneuver one another in transit. Spatial arrangements that engender statelessness signal a new strategy adopted by authorities to control undocumented migration and asylum flows.

Steven Spielberg directed a 2004 film titled *The Terminal*, in which a man, played by Tom Hanks, is lost in limbo in New York's John F. Kennedy International Airport. Hanks is denied entry to the United States because his visa, from a country that ceased to exist while he was in the air, is no longer valid. Marc Augé (1995) characterizes international airport terminals as the "non-places" of supermodernity. Stateless in the non-place, Hanks makes this ambiguous space his home for nine months. With *The Terminal*, Hollywood brought the drama of legally ambiguous spaces to popular culture. But long before the film's release, the idea of stateless zones resonated with even the most casual observers of current events. Since the Halliburton Corporation constructed Camp X-Ray, a detention center on the United States' naval base in Guantánamo Bay, Cuba, stateless space has been a major part of public discourse in the United States and internationally. Camp X-Ray was built and utilized to house people detained during the retaliatory U.S. military action in Afghanistan, carried out to suppress the Taliban and the al-Qaeda network responsible for the September 2001 terrorist attacks in New York and Washington, D.C.

Both the film and the ongoing debates about the status and legal rights of prisoners in Guantánamo Bay highlight the arguments developed in this chapter. Nation-states are increasingly practicing enforcement in transnational, dispersed fashion. Part of this dispersion involves the creation and exploitation of geographically and legally ambiguous sites. Like the long tunnel, these are sites of paradoxical sovereignty (Agamben 1998): at once inside and outside of sovereign territory, at once characterized by both the presence and the absence of nation-states. Just as the United States government refused to move detainees at the Guantánamo base to sovereign territory where they would access a broader range of legal rights, states keep asylum seekers detained in locations where their access to asylum and legal representation remains limited. Spatial arrangements that render people stateless by geographical design are becoming an enforcement tactic to control people on the move who fall into the categories of undocumented, unauthorized, or "irregular" migrants. These populations constitute what policymakers would call "mixed flows"—some moving in order to work and others searching for protection from persecution.

Policing beyond sovereign territory illustrates the expansion of national boundaries into ambiguous zones of sovereignty. Many state practices initiate and enable transnational border enforcement, including increases in transit visas and other excessive requirements for documentation; increased investment in human resources to support interdiction abroad; increased "front-end" security procedures such as background checks; and the integration of intelligence, border enforcement, and other security-related industries. Other practices include economic sanctions for private carriers that transport "irregular" migrants; Safe Third Country agreements like the one implemented between Canada and the United States in 2004 (discussed in chapter 3); "protection in region of origin" that keeps refugees in camps closer to home; and movement of asylum processing farther afield, undertaken especially by the United States and Australia. The move toward regionalization in refugee management means that more efforts are being made to house and process refugees in camps in their regions of origin (see Castles and Van Hear 2005; Schuster 2005b).

One mechanism that maintains migrants at arm's length from sovereign territory is the creation of sites that render people stateless by geographical design. To illustrate this phenomenon, I discuss Australia, North America, and the European Union, those regions that have sought most aggressively to police borders abroad through remote immigration control, introduced in chapter 2, using tactics developed to stop transnational migrants from reaching sovereign territory.

Exploration of these zones demonstrates an ethnography of the state that is a matter of concern for refugee advocates and that shows the state turning itself inside out through daily practices of enforcement both on- and offshore. In this chapter, I develop a typology of four kinds of sites: remote detention centers within sovereign territory that restrict access to refugee determination processes, detention facilities offshore, short-term stateless zones associated with transit, and dynamic sites of interdiction abroad such as those in Hong Kong discussed in chapter 2.

Faced with highly visible smuggling in 1999, the Canadian government took advantage of the microgeography of the Esquimalt base to place migrants in a more ambiguous locale, where their legal status came into question for a period of time. Canada was not alone in pursuing the "long tunnel" strategy. It was in the context of the trend toward stateless zones evident in Australia and the United States that the Canadian government successfully designated the naval base where migrants were being processed as an area that was "not-Canada" (see Perera 2002c). This episode provides a window into these creative geographies of exclusion. The transition of the base from sovereign to stateless space corresponded with more aggressive strategies elsewhere. Australia and the United States were practicing interdiction on the high seas and diverting ships to territorial and nonterritorial islands to prevent migrants from reaching their shores. As states move detention, interdiction, and the processing of asylum applicants farther afield, it becomes more difficult for undocumented migrants and potential refugee claimants to reach sovereign territory. Instead, people in transit increasingly find themselves in locations where their legal status is unclear and their access to refugee programs is limited.

As nation-states use geography strategically to deny access to asylum, voids in policymaking create conditions for crisis. Securitization trumps protection with the enactment of exclusionary measures in extraterritorial locales. These practices label, homogenize, hide, and contain transnational migrants in particular ways. These are practices enabled by public discourse that criminalizes asylum seekers as threats to security.

Civil servants collaborate across international borders and look to one another's strategies to develop and enact enforcement strategies. Because states look to one another in developing enforcement strategies, it is essential to broaden the analytical picture beyond individual nation-states to include converging enforcement patterns in the international context. A global pattern of formal and informal enforcement practices emerges.

Shifting Geographies of Enforcement

The post-9/11 security environment has seen not only major legislative changes, but also major shifts in the governance of migration and, specifically, its connection to securitization (Bigo 2002; Sparke 2005; Amoore 2006; Huysmans 2006; Coleman 2007; Amoore and de Goede 2008). In the United States this entailed the creation of the largest federal bureaucracy in 2003, the Department of Homeland Security, which took over all immigration functions, along with many others. Canada pursued a distinct strategy and formalized the "division between church and state," as one respondent called it, by separating the enforcement mandate from facilitation with the creation of the Canadian Border Services Agency in 2003. While member states of the European Union continued to struggle over the harmonization of asylum and immigration policies, they collaborated in 2005 to create Frontex, a united agency to police the perimeter of "Fortress Europe."

Areas that are stateless by geographical design require a remapping of the boundaries of sovereign territory. These areas are one link in a chain of enforcement mechanisms that are shrinking the spaces of asylum. Paradoxical zones along the margins serve as a dramatic expression of changing practices of sovereignty, and as the places of encounter between states and transnational migrants. Through processes of "externalization" (Betts 2004; Castles and van Hear 2005), as they are designated in the European context, states turn themselves inside out. If not stopped through interdiction at sea or in foreign airports, undocumented migrants are "turned back" in domestic airports, or detained remotely if they succeed in reaching sovereign territory. Symbiotically, as enforcement mechanisms grow more aggressive vis-à-vis the transnational state, so too does the human smuggling industry.

Juxtaposing the work of civil servants and the work of human smugglers calls attention not only to the *macrogeography* of global flows from one region to another, but also to key *microgeographies* of local encounters. The dynamic geographies of the naval base and the international airports where smugglers operate are key cases in point. Civil servants have limited resources and face constraints of law and jurisdiction, but they compensate by sharing information, mapping trends, and operating strategically in "hot spots" where smuggling originates and transits.

As states choose from a variety of strategies to combat human smuggling, detailed in chapter 2, they have tended to increase "front-end controls" enhanced by a more securitized climate. For a while these seemed to be quiet geographies of enforcement enacted in daily practice, forming a patchwork quilt that constrained those on the move, often operating beyond the radar of refugee advocates and activists. More recently, these sites have become more visible and therefore more contested in public discourse. I fear that they are becoming the norm, and that it was in the context of aggressive interdiction and detention strategies undertaken by Australia, the United States, and European states that the rescripting of the base transpired in Canada.

In fact, states are proving themselves to be as effective as human smugglers and drug cartels at building tunnels. According to Vice President Dick Cheney, speaking on *Meet the Press*, "A lot of what needs to be done here will have to be done quietly, without any discussion, using sources and methods that are available to our intelligence agencies . . . that's the world these folks operate in. And it's going to be vital for us to use any means at our disposal, basically, to achieve our objective" (Mayer 2005). Cheney identified this as "work[ing] through, sort of, the dark side."

Cheney's statement and the accompanying practices prompt urgent questions. *Where* is the dark side? What goes on there? How do states manage to subvert and circumvent the law there? Do they have limitless capacity to manipulate geography with the construction of long tunnels and the movement of ports of entry? Is this "no man's land," "a third category [that] cast them outside the law," according to international law experts speaking on detainees at Guantánamo Bay (Mayer 2005)? Or is it *some* man's land? And if so, whose?

For asylum seekers, the dark side involves the threshold between those sites mapped formally as sovereign territory and those mapped informally around, over, under, or in subversion of the law. To return to de Certeau's aphorism, "What the map cuts up, the story cuts across" (Conquergood 2002, 145), the threshold with the dark side is detention and its dispersal of migrants in remote locations (e.g., Bloch and Schuster 2005; Schuster 2005a). These issues raise questions about the integrity of signatory states regarding their commitments to give asylum seekers due process (Brouwer 2003).

Remote geographies of detention often involve the activation of residually militarized island landscapes. The United States detains "illegal enemy combatants" in Guantánamo Bay and other "extraterritorial" locales. Torture is outsourced through extraordinary rendition, the export of torture and detention to countries where it is not prohibited. Australia, Canada, the United Kingdom, the United States, and New Zealand develop more aggressive practices of interdiction abroad where Airline Liaison Officers and Immigration Control Officers operate informally to stop potential refugee claimants in locales where they hold no jurisdiction.

Australia's "Pacific Solution" involves the "power of excision" to declare islands such as Christmas Island and other places where ships land with migrants—including the coast of mainland Australia—retroactively no longer part of Australia (see Map 3). Borders are thus pushed farther away and nearly erased in this case, where asylum processing is contracted out to poorer countries. This combines with remote detention (inside sovereign territory) to create a powerful geography of exclusion.

Meanwhile, the UK, Germany, and Italy have each put forth proposals to send all asylum seekers to processing centers in Tunisia, Morocco, Libya, Algeria, and Mauritania (see Schuster 2005b). While protests by other EU members slowed these plans, Italy moved forward in collaboration with returns to Libya (European Parliamentary Group 2006). Still other migrants attempting to reach the EU are held in detention centers in Ukraine (Human Rights Watch 2005).

Enforcement is becoming more transnational, informal, and privatized (Flynn and Cannon 2009). Information is shared easily, and those managing detention may not be civil servants at all, but rather third parties such

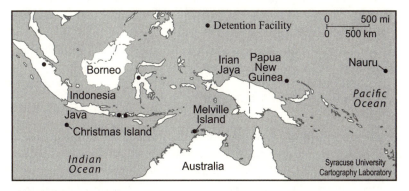

Map 3. Australia's Pacific Solution, in which migrants' ships were intercepted and migrants brought to detention facilities offshore. Some of these sites are on Australian soil (e.g., Christmas Island), whereas others are independent nation-states to which Australia contracts out detention (e.g., Nauru, Indonesia).

as private corporations, international organizations (such as the International Organization for Migration), or local authorities where offshore detention facilities are located. Border enforcement is transnationalized, integrated among nations, and privatized with large government contracts to corporations such as Accenture that create smart cards and biometrics to support greater interdiction. Spain has designed an electronic barrier along its coast closest to Morocco. Detention practices are contracted to private companies, including Halliburton, Wackenhut, and GSL, that manage centers in Australia and the United Kingdom (Flynn and Cannon 2009).

Worldwide, signatory states are successfully undermining their responsibilities to the 1951 Convention Relating to the Status of Refugees, as well as its 1967 Protocol, through enforcement practices enacted extraterritorially. They are even contradicting their own national laws, as in the case of Australia's decision to excise its own territory for the purposes of migration. This means that asylum seekers landing on Australian territory do not have the rights accorded to an asylum seeker by law because Australia has excised its sovereign territory for the purposes of its own migration law. All asylum seekers will be processed "offshore" on Christmas Island (Australian territory), Nauru (an independent country), or elsewhere. This remarkable exceptionalism illustrates how the intersection between geography and law is manipulated to stitch together an archipelago of dispersed sites of enforcement and detention, where people are rendered stateless, unable to access sovereign territory and asylum systems. Their legal status is ambiguous, as are opportunities for advocacy and appealing of decisions.

The nation-states practicing these aggressive exclusionary measures—Australia, the United States, and Canada in particular—are those with the highest rates of refugee resettlement per capita and the most pronounced anti-immigrant public discourse, where poor people on the move between states are criminalized and securitized.

Stateless by Geographical Design: A Typology[3]

Enforcement activities that transpire within sovereign territory correspond with a number of important extraterritorial activities. The naval base where migrants were processed in BC is part of a broader pattern of intensive interdiction being practiced by nation-states, part of a pronounced global trend toward the creative design of sites of detention and enforcement. The designation of the base as a non-Canadian space corresponds with increases in stateless rooms in airports and the remote placement of detention centers off the coasts of Australia and the United States.

Just as migrants frequently change the routes and methods of their journeys, so do states alter the geography of enforcement. Immigrant-receiving nations such as the United States, Australia, the United Kingdom, and Canada create stateless spaces as a means for coping with what they call "irregular migration." While not new, these practices have been fueled recently by security discourse following the terrorist attacks of 2001.

Receiving states are changing the geography of their policing strategies through their transnationalization. Some tactics are more reactive, including detention and expedited removal (McBride 1999, 304). Other methods are more proactive, such as diverting ships carrying migrants to nonterritorial islands, creating stricter requirements for transit visas, imposing carrier sanctions on commercial companies that transport migrants with false documents, training airport and border personnel to profile migrants, strengthening border security systems, entering "safe third country" agreements (Macklin 2003b; Hyndman 2004), and implementing protection in regions of origin (see Canadian Council for Refugees 2003, 3). The strategic use of geography to suppress smuggling and to diminish national responsibilities proves central to each of these enforcement tactics.

These zones come into being when governments take advantage of the conditions of crisis. They activate networks of civil servants, ease the sharing of information, alter legislation, and take transnational measures in the name of national security. Too many ad hoc decisions about migrants and refugees are made (Massey et al. 2002), and stateless spaces may well be an outgrowth of this reactive practice of policymaking. In the current security climate, states are now as concerned, as contradictory, and as creative as ever in their desire to facilitate some migrations and suppress others. The sites that they create are zones of ambiguity, rendering those intercepted and detained stateless by geographical design. It is essential to understand where and how these places come into being, who ends up there, and why. Hardly Augé's "non-places," these sites come into being in historically significant and strategic ways to intervene in transnational mobility.

In these shifting institutional landscapes, some sites exist only temporarily, while others become more permanent. Here I discuss four kinds of zones that render migrants stateless by geographical design. Governments are pursuing more remote detention practices (inside and outside of sovereign territory) that place detainees in locations where they have restricted access to refugee programs. They are also designing dynamic zones where interdiction and short-term detention take place on airplanes and in airports where one is not yet landed in sovereign territory. Finally, states are stepping up interdiction practices by sending civil servants abroad to informally police borders and suppress undocumented migration. I am concerned about what is happening to people on the move with a desire to work or make a refugee claim in another country, and I review each type next.[4]

Remote Detention Paradoxically "Internal" to Sovereign Territory

Observation of institutional actors involved in the Canadian response to human smuggling reveals tension between two narratives of immigration policies pertaining to smuggled migrants and refugee claimants. The first narrative was that of the powerful state, able to intercept, detain, and repatriate. But the second narrative, recounted by civil servants, was one of powerlessness, of marine arrivals generating crises for underresourced bureaucracies. These modes of crisis actually paved the way forward for the federal department. The power of immigration policy tends to come from its regulation (Macklin 2003a, 474–475), and policy toward human smuggling is no exception. By neither designing long-term policies nor allocating adequate resources to plan and prepare for marine arrivals, states structure the conditions for de facto crises.

States collude in the production of crises. As they unfold, they enable an avoidance of planning and a refusal to grapple with "the big picture."[5] What emerges to fill such voids in policy are reactive measures implemented during times of crisis, known as "policy on the fly" in Canada and "policy on the run" in Australia. Policies on the run give rise to long tunnels.

The detached geography of detention in Canada, outlined in chapter 4, resonates with even more remote detention practices undertaken by the United States, Australia, and—increasingly—EU member states. The media images of the bogus refugee were central to the project of constructing statehood and were part of the explanatory narrative enabling the manipulation of legal status to control offshore flows—images corresponding to Bonnie Honig's (2001) foreigner to democracy and Agamben's (1998) "bare life." As migrants move globally between economically disparate regions, antiimmigrant and antirefugee sentiments sharpen in public dialogue in the places to which they move. The popular media contributes greatly to the blurring and criminalization of categories of migrants and refugees. This discursive rise of the bogus refugee is associated with more exclusionary measures and, therefore, a dramatic drop in the number of refugee claimants. An important relationship exists between public discourse about migrants and their geographic locations.

Remote detention takes different forms in different regions and states. These remote geographies of detention attempt to remove asylum seekers from urban centers where support and advocacy communities are located. Esquimalt and Prince George, where the 1999 migrants were processed in British Columbia, serve as more subtle examples of the move in this direction.

Australia offers a more pronounced geography of spatial isolation. Following a policy of mandatory detention implemented in the early 1990s, asylum seekers were detained in remote locations along the northwestern coast, such as Darwin, and in the outback, in Woomera, for example. These centers were located far away from larger centers of population offering resources that would assist people to access asylum: interpreters, legal representation, diasporic communities, and advocates. As detention centers in more remote locations have been shut down in recent years, detainees were moved to centers located in larger cities, where they were could be more connected to urban populations and resources, such as the Villawood Detention Centre located in the suburbs of Sydney and depicted in Figure 10.

In the United States, detached geographies of detention take yet another form, with dispersion by mixing immigrant and asylum-seeking detainees in with "mainstream" prison populations (Welch 2002). Less than 50 percent of detainees are held in dedicated federal detention facilities that hold only persons with foreign citizenship; many others are incarcerated among mainstream populations in county jails that are not dedicated to migrant detention. Migrants are arrested in various locations throughout the country and funneled to remote rural county prisons and large, dedicated federal facilities, many located in the south. They are moved quickly and frequently between detention centers.

Dispersal across regional contexts in the form of movement through the system to remote sites makes location, identification, and communication with detainees difficult and obscures statistical accounting of the numbers of foreigners being held in detention by authorities at any given time. In both Australia and the United States, people are moved quickly, overnight, without warning to the detainees, their families, or their lawyers. Not only does this make it difficult for families to find them, but networks of advocacy must be restored or created anew.

Social movements organized around antidetention and antideportation have grown in cities and in rural areas alike to protest the isolation

Figure 10. The Villawood Immigration Detention Centre on the western outskirts of Sydney. Some asylum seekers were detained for years at Villawood as they awaited the resolution of their cases. Unlike those migrants intercepted at sea and detained on islands, these detainees made it to sovereign territory and were able to make an asylum claim. Some among them will become stateless while in the detention center and, even once released, denied asylum in Australia but unable to return to their home countries. (Photograph by author.)

of remote detention. Tightly networked advocacy communities share information about the location and status of detainees. Their activities included mechanisms to welcome, reconnect with, and support the release of detainees through letters, visits, and legal and psychosocial support (Lonely Planet 2003). Even where migrants were detained offshore, activists and advocates worked to pull them into their challenges to those entities onshore that they held accountable for the offshore activities.

Offshore Detention and Processing

While the episodes above transpired *internally* to sovereign territory, the United States and Australia implemented even more aggressive measures— detaining and processing asylum seekers far offshore, remote from mainland territory, in areas where smuggled migrants making refugee claims have even less access to due process. In some cases, they essentially contract out signatory responsibilities to private companies, to other countries, or to the International Organization for Migration (IOM). Hence, detention practices abroad work in tandem with more expansive detention practices "at home," particularly in the case of the United States (Simon 1998; Welch 2002) and Australia (Bashford and Strange 2002; Perera 2002b).

The most prominent example of stateless detention in contemporary public debate involves "enemy combatants" held at Camp X-Ray in Guantánamo Bay, Cuba. There, on another base, the United States holds detainees in a remote location removed from embassies and advocates, where the legal status for these detainees remains unclear. Guantánamo Bay has also been utilized for years by the United States to keep Haitian and Cuban migrants in legal and spatial limbo on the return route following marine interceptions (Crisp 1997; McBride 1999). Here the United States exploits an uncertain zone between states to evade the obligations of international humanitarian law, a strategy that renders many of those detained on the military base stateless by geographical design. Asylum seekers detained in ambiguous sites between states are in a legal limbo, where political status of location intersects with legal status of the individual. Due to the ambiguity of location, they are neither asylum claimants nor immigrants; they cannot usually work, nor do they have the right to leave. In some cases, they cannot be returned home. They find themselves trapped temporally and spatially.

The United States has also detained asylum seekers on the Pacific islands of Guam and Tinian (United States Committee for Refugees 1999). These two islands hold distinct political statuses in relation to the United States, and therefore asylum applicants on each island face differing restrictions on their mobility. While migrants can file claims on Guam, an unincorporated U.S. territory, they may not travel to the mainland United States unless and until they are accepted as asylum applicants. On Tinian, on the other hand, U.S. immigration laws do not apply.[6] In early 1999, when large numbers of Chinese migrants were arriving on Guam, detention centers were operating beyond capacity, with makeshift tents housing asylum applicants and few immigration lawyers available. The U.S. Coast Guard began to divert boats headed for Guam to Tinian instead (United States Committee for Refugees 1999). Known as the Tinian Operations (United States Committee for Refugees 1999, 6), this series of negotiations on land and water between smugglers and state authorities is reminiscent of the game metaphors through which Okólski (2000) and Andreas (2000) describe human smuggling and border enforcement. Borders are never policed strictly along the lines on a map (Andreas 2000; Salter 2004; Koslowski 2005).

As the number of asylum seekers in Australia peaked in the mid- to late 1990s, restrictive measures there reached a pinnacle with an interception incident in August 2001. Following the controversial *Tampa* incident in September 2001, when Prime Minister John Howard's government would not allow 433 migrants rescued from a sinking ship to disembark on sovereign territory, Australia created the "Pacific Solution" (Hugo 2001; Mares 2002). Migrants rescued by the Norwegian *MV Tampa* were detained for many months on the tiny, poverty-stricken island nation of Nauru, where the IOM managed detention, and on Manus, a territorial island of Papua New Guinea. They were also processed on Christmas Island, a small territory of Australia close to Indonesia (Marr 2009), where in 2006 Australia built a new detention center to house 800 to 1,200 migrants. A small number of migrants from the *Tampa* were eventually resettled in Australia, but the United Nations High Commissioner for Refugees (UNHCR) negotiated the resettlement of most in "third countries," including Canada and New Zealand.

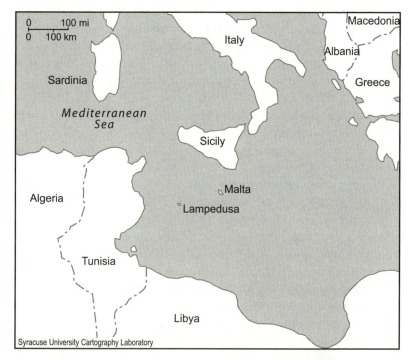

Map 4. Italy's Lampedusa, an island, where thousands of migrants traveling from Africa by sea were detained. The reception center has been the subject of much controversy surrounding the detention of asylum seekers, deportation, and human rights abuses in detention.

Australia subsequently changed the status of its island territories in an emergency meeting of Parliament in order to prevent asylum seekers who had landed by boat from being able to access its refugee program. This "solution" entailed the enactment of restrictive legislation, marine barricades to prevent ships from reaching Australian shores, contracting out processing to countries such as Nauru and Papua New Guinea, and the use of the "power of excision" (see Hugo 2001; Perera 2002a). This power enables the state to declare sites where migrants have landed to be suddenly no longer a part of sovereign Australian territory.

In 2005 Italy joined ranks with Australia and the United States, deporting large groups of migrants from the small Italian island of Lampedusa (see Map 4), near Tunisia, to Libya, against the wishes of the UNHCR (UNHCR 2005).

Until 2004 "clandestine" migrants—as they are called locally—who landed on Lampedusa and made a claim would be transferred to reception centers in Sicily for processing. Italy entered into bilateral arrangements with Libya that involved construction of detention centers, policing of the Mediterranean, and facilitated returns from Lampedusa to Libya and from Libya to Eritrea (European Parliamentary Group 2006).

In response to the group deportations from Lampedusa to Libya, a country that is not a signatory to the 1951 Convention Relating to the Status of Refugees, UNHCR spokesperson Ron Redmond said, "With a capacity of a mere 190, the centre is easily overwhelmed, creating an air of crisis that is perhaps not strictly necessary." With some 11,000 claims, Italy receives among the fewest asylum seekers annually of the larger EU member states. But like Australia, the United Kingdom, Canada, and the United States, Italy has crafted long tunnels where enforcement takes place, particularly along its southern border, where most detention centers are located (see Rutvica 2006).

Farther west in the Mediterranean, Spain has also confronted a dramatic increase in West African migrants arriving in fishing boats on its Canary Islands in 2006. As "Fortress Europe" expands, both economic migrants and asylum seekers arrive at its outermost edges seeking entry. In response, EU member states have enhanced their Frontex policing operations in the Mediterranean.

Many of these practices occur on small islands, sometimes controlled or funded by larger, wealthier states like the UK and Australia, with island mentalities. Australian authorities, for example, boast of the strongest border compliance practices in the world (Interview, Canberra, April 2006). These policies include requirements that all visitors to Australia obtain a visa and that all asylum seekers face mandatory detention, a harsh deterrence strategy for those seeking protection.

The UK, too, has exercised aggressive moves farther afield to distance asylum seekers from its shores. It enacted the Nationality, Immigration, and Asylum Act in 2002, which restricted support for asylum applicants and provided for processing in detention centers (United States Committee for Refugees 2003). Shortly thereafter, the notorious White Paper, leaked from Prime Minister Tony Blair's government, proposed the processing and detention of all asylum seekers in "regional protection zones," closer to source and transit sites, and farther away from sovereign territory. Proposed locations for these "international asylum centers" included Albania, Croatia, Iran, Morocco, northern Somalia, Romania, Russia, Turkey, and Ukraine (Schuster 2005b). Human rights groups have expressed concern that this strategy undermines the integrity of the 1951 Convention and places vulnerable populations in countries where they are in greater danger of having their human rights violated by police misconduct and by poor detention and arrest practices that include torture (Human Rights Watch 2003).

Interdiction

In recent years, Canada, the United States, the UK, Australia, and New Zealand have all increased interdiction abroad. Interdiction is a legal term for "disrupting" journeys en route. Interdiction happens in airports and at sea. The United States, for example, is notorious for boarding boats in international waters, well beyond its territorial jurisdiction. Australia, too, has led the way at sea vis-à-vis its "Pacific Solution," which involves aggressive interception practices at sea and the most extensive set of bilateral arrangements between states for the return of migrants. It entails routine interception of boats carrying migrants and preventing them from reaching sovereign territory. The migrants are returned to where they came from, detained in Indonesia through bilateral arrangements, or moved to other island territories that are either Australian territories or "safe" independent countries, essentially outsourcing asylum responsibilities.

These interception practices corresponded with the location of detention sites on islands distant from mainland territory to which people intercepted at sea by the Australian navy were towed. Whether detention took place internally or offshore, both strategies placed migrants in isolated areas where there was little infrastructure for legal advocacy (Hugo 2001, 31; Bowden 2003) and illustrated a concerted effort to exclude migrants from sovereign space.

Another "front-end" interdiction strategy that stops migrants long before they reach sovereign territory is that of deploying civil servants to police distant, "preborder" locales. Audrey Macklin has called Canada "something of a pioneer in instruments of interdiction" (2003b, 5) with its creation of the "multiple borders strategy." It has since led other countries in increasing its numbers of Immigration Control Officers, Migration Integrity Specialists, and Airline Liaison Officers working on the ground abroad in hot spots such as Hong Kong, mainland China, and in other countries in Asia and Africa. This strategy was developed in the 1980s. Canadian Immigration Control Officers have been working in this capacity abroad since then, but have increased in number with more resources for security since 2001. In 2003 Canada had eighty-eight such officials practicing enforcement abroad, the UK had thirty-six, and Australia had fifteen.

These officers have little to no jurisdiction, but enact an informal web of security by feeding information on migration to authorities in the host countries where they are located and by training airline employees to look for false documents and suspicious persons. With a mandate to detect undocumented migrants and deter them from reaching shore, these civil servants act informally as liaisons between foreign embassies, private security companies at airports, airlines, and host authorities. They share information in order to suppress smuggling and prevent people from reaching sovereign territories to make refugee claims (Citizenship and Immigration Canada 2001). One of their primary goals is to look for—and to train airline employees to look for—fraudulent documents, which have been on the rise since the implementation of stricter visa requirements (Brouwer 2003).

Known as "front-end controls," these measures work in tandem with sanctions implemented to punish carriers who transport persons with false documents. These front-end controls have prompted a number of scholars to reconceptualize the policing of borders beyond the usual sovereign boundaries (e.g., Lahav 1998; Lahav and Guiraudon 2000; Hyndman and Mountz 2006; Coleman 2007). Front-end tactics contrast with the more reactionary practices of detention and deportation of migrants and exemplify yet another form of the privatization of enforcement (see Lahav 1998). Not only are private companies, such as airlines and transport industries being paid *by* states to build and run detention centers and police borders, but they are also forced to *pay* states when they fail.

Stateless Zones in Airports

These moves toward more remote detention and more aggressive interdiction coincide with the growth of stateless spaces in airports. Stateless rooms are cropping up all over Europe as frontline immigration officers isolate migrants either before they disembark from planes, by demanding their documents on the aircraft, or once they have entered the airport (Figure 11). On a global scale, those seeking asylum and those seeking work are increasingly finding themselves lost in this new policy—in Augé's non-places (1995), the spaces between states.

In London's Heathrow International Airport, Paris's Charles de Gaulle International Airport, and smaller airports across Europe, civil servants from multiple countries work the "international zones." Like the military bases, these zones are not France, not the UK, but rather ambiguous sites that are neither here nor there.

Gillian Fuller discusses the Australian airport to highlight the ambiguity of these sites:

> The airport . . . involves the incorporeal transformation of the traveling body—as a citizen, a passenger (pax), a baggage allowance, an accused, or an innocent. The airport constitutes a space where a series of contractual declarations (I am Australian, I have nothing to declare, I packed these bags myself) accumulate into a password where I am free to deterritorialize on a literal level. . . . When citizenship of a mass-transit world entails neither blood (born of citizen parents) nor soil (born in sovereign territory), as in,

Figure 11. Another long tunnel, this one en route to the processing of arrivals on a China Air flight landed at Vancouver International Airport. (Photograph by Jennifer Hyndman.)

say, "multicultural" Australia, "the continuity of man and citizen, nativity and nationality" (Agamben 1998, 131) is broken—and with it some of the fundamental presuppositions of modern sovereignty. . . . The camp, like the airport, is built for transit. Yet in the camp, no one moves. Both airport and camp constitute zones of exception, each . . . framed by a rhetoric of emergency. (2003, 4–6)

Also drawing on the work of Giorgio Agamben (1998), Suvendrini Perera (2002c) links the detention center, which she labels "not Australia," with Camp X-Ray in Guantánamo Bay, and cites both as examples of Agamben's camp. This exercise of sovereignty places the bodies of migrants and civil servants alike in ambiguous, stateless locales that are simultaneously inside and out. Agamben calls this the crisis of sovereignty (1998). Perera (2002b) argues that for imprisoned asylum seekers, war abroad becomes war at home. I would extend Perera's argument to include not only the paradoxical spaces such as the Woomera detention center (now closed, but another former military base) operating *within* sovereign territory, but also those stateless zones operating *beyond* sovereign territory, where people become stateless by geographical design. There, for the state, the war against undocumented migration at home becomes the war abroad.

With each of these sites, one is left with the image of a rather powerful, proactive, transnational set of state practices wherein civil servants extend the web of enforcement with each step farther away from sovereign territory. Together, they form the multisited, transnational constellation of civil servants linked through intelligence and enforcement networks (Salter 2008). By isolating migrants in remote locations, states restrict access to territories where they might make refugee claims or take up residency illegally. These sites prompt legal ambiguities and raise questions about where states (and state actors, more specifically) locate themselves, where they locate borders, and where they exercise power to position migrants as *homo sacer*, bare life, stripped of rights. When American civil servants police international airports abroad through their Global Outreach program, for example, the borders of the United States are extended beyond their traditional continental perimeter. While the number of civil servants at work in these fields may not be large in comparison to the number of undocumented migrants on the move, their impact is significant and their success celebrated by federal agencies (Citizenship and Immigration Canada 2001). In a recursive relationship, states concretize enforcement, migrants employ smugglers, smugglers study the geography of states, and states study the geography of the smugglers.

The innovative geographies of stateless sites and interdiction abroad illustrate the truly transnational and increasingly dispersed management and relocation of national borders. Local struggles are linked to those occurring at a global scale. The seemingly ad hoc nature of designating ambiguous zones contrasts with the frequency with which they are appearing, and suggests a typology of these spaces that can serve as a foundation for their comparative study.

The typology could be further aggregated by stateless spaces that operate within sovereign territory—such as the Esquimalt naval base on Vancouver Island, remote detention in the Australian outback, and stateless spaces in airports—and stateless spaces at work quite literally between sovereign territories—such as interceptions in international waters on the high seas, detention in Guantánamo Bay, and the outsourcing of asylum responsibilities to Nauru and Papua New Guinea. The testing and refinement of this typology may challenge the very nature of sovereignty that social scientists often accept as given: as contained within a compact territory. It is not clear to prisoners, to their legal counsel, or to the American courts precisely what laws and human rights protocols apply to paradoxical sites such as Guantánamo Bay. The same is true for asylum seekers detained in centers in "third countries." It is not clear, for example, who is responsible or accountable for detainees being held in centers run by the IOM on behalf of Australia on Nauru and Lombok. Where are these sites? Their locations can be mapped, but the legal geographies attached to them remain ambiguous, leaving advocates with a legal conundrum in determining how national and international laws apply (Goodwin-Gill 1986; Paucurar 2003). Such ambiguities affirm the need to study the microgeographies of sovereignty. What happens on airplanes, in stateless rooms in airports, and on container ships influences whether sites will be treated as sovereign locales and what rights are available there, including access to asylum and legal representation.

This typology maps and categorizes sites that are stateless by geographical design and questions why, where, and how these places operate. Resorting to these sites perpetuates the cat-and-mouse game between smugglers and officials (see Okólski 2000). The sites demonstrate states' ability to alter time and space and offer an understanding of the shifting relationships between migrants, refugees, and nations. It is significant that this trend toward stateless spaces converges with migration between economically disparate regions of the world. Anti-immigrant and antirefugee sentiments sharpen in public discourse in the United States, the UK, and Australia through the criminalization of migrants in popular media. The wealthier countries tighten control, and asylum seekers are increasingly distanced. Strategies that at first appear ad hoc and far-flung, when stitched together establish a more concerted effort to exclude.

This chapter has discussed examples of stateless spaces related to Canada, the United States, Australia, and the United Kingdom. These four immigrant-receiving countries are leaders in enforcement measures intended to control unauthorized migration. They have a number of other significant characteristics in common, including extensive coastlines, advanced economies, commonwealth or former colony status (with the exception of the UK), histories of migration and settlement, and multicultural populations. Also, they all exercise sophisticated border control policies. The shift toward stateless spaces is taking place globally and is quickly becoming a major tool that many governments use to manage mobility and reduce numbers of asylum seekers. As this preliminary typology evolves, it should differentiate between the kinds of nation-states pursuing such strategies, perhaps by geography, demography, or governance strategies.[7]

The shifting governance strategies outlined herein raise important questions regarding the legality, sustainability, and efficiency of these policies for deterring illicit migration on a long-term basis. These tactics affect not only those moving illegally, but also refugees who are attempting to make claims (Brouwer 2003). By pursuing interdiction abroad, nations may be in violation of the principle of *non-refoulement* and evading their responsibilities as signatories of the 1951 Convention and its 1967 Protocol (Macklin 2003b). As nations attempt to balance efforts to facilitate immigration and enforce borders, the categories of smuggling, refugee claimants, and illegal migrants increasingly run together. The terms that govern refugees, smuggled migrants, and terrorists may now be so blurred as to render obsolete the national refugee policies and international protocols developed to categorize and govern them.

Resubjectification and Exception

How are we to conceptualize zones that are stateless by geographical design? Many scholars have turned to Giorgio Agamben's (1998) text (see Mitchell 2006) because he too dwells in tunnels and at thresholds between states. Agamben spatializes the process of subjectification wherein migrant and sovereign identities are produced through processes of exclusion. In his typology, states and imprisoned subjects "illuminate" one another. His analysis centers on the creation of the camp as an exceptional site engendered by crisis that ultimately becomes normalized. He sees paradoxical "zones of indistinction"—where it remains unclear whether one is in or out—as emblematic of the power, essence, and spatial arrangements of sovereignty. The state externalizes, internalizes, and thrives on these "excessive" zones:

> The sovereign exception is the fundamental localization, which does not limit itself to distinguishing what is inside from what is outside but instead traces a threshold (the state of exception) between the two, on the basis of which outside and inside, the normal situation and chaos, enter into those complex topological relations that make the validity of the juridical order possible. (Agamben 1998, 19)

The dispersed detentions discussed in this chapter are born of crisis. Through their creation, state and imprisoned subject remake one another and realize the correspondence between geography and discourse. When one is detained, the process of subjectification intensifies through geography; the person's *identity* is now conveyed through *location*. The remoteness of detention allows a degree of concealment; authorities do not release information or identities, or permit human rights monitors or advocates. As much as the nation-state tells us something about the detainee through imprisonment and crisis enactment, the detainee (detained in sovereign and extraterritorial locales) relays the changing practices and spatial arrangements of the state, particularly in the marginal and paradoxical zones where people are included through exclusion.

The association between transnational migration and criminality has been reinforced by the media and linked to antiterrorism legislation that privileges the security of states and their citizenry over the security and protection of "foreign nationals." In the conflation of public discourse about terrorists, refugees, economic migrants, human smuggling, and others on the move, people are stripped of their identities as individuals and lumped into groups, as in the case of the refugee claimants in BC.

Though boats organized by smugglers are known to carry "mixed flows," refugee claimants among them are scripted as economic migrants and perceived as bogus refugees before the adjudication of their claims. Such labels have racialized and criminalized this group in relation to the nation-state and saturated the media. The explanatory narratives suggested that these claimants are not "Convention refugees." Institutional geographies emerge and are manipulated retrospectively to fit the explanatory narrative, exemplified by those created in BC and discussed in chapter 4.

Agamben's narrative is an easy one to apply to this exclusion, and yet, as scholars have argued, there are other stories to tell (Sanchez 2004;

Pratt 2005; Mitchell 2006). One way to resist Agamben's totalizing narrative, in which we all perch precariously outside the state, is to come to understand the intimacies of exclusion where performative states redefine borders at the scale of the body. The daily encounters and negotiations between nations and migrants enable conceptualization of subjectivities of each through the other. Each makes and remakes the other in the daily spaces of these contact zones, where multiple processes transpire between state and body. Bodies are differentiated while simultaneously grouped, homogenized, racialized, medicalized, and criminalized. Containment and homogenization of identities enable the correspondence between geography and discourse. Australia, for example, provides detailed statistical portraits of where people are detained but no details on demographics such as country of origin. The United States, too, hides information by detaining migrants among mainstream populations in county jails and not releasing demographic information on foreign-born detainees. Identification must be provided by those wishing to visit detainees, yet no identification of those arrested and detained is released.

Like the Chinese migrants detained and ultimately deported from British Columbia, those detained in remote locales between states do not count and are banned from becoming refugees through preclusion of their rights. They do not have individual identities but are instead a collective threat, which explains why they are *there*, safely at a distance from here. These are among the violent acts committed against people in these locales, where violence passes over into law in Agambennian terms. By being prevented from reaching sovereign territory, people on the move are forbidden from ever joining the juridical order. In Agamben's parlance, they are included through exclusion. They must fight to be named and recognized by the sovereign administrative power in the territory where they had struggled to land.

Civil servants experience crisis in the response to human smuggling. But why would state institutions designed to deal with international migration—including human smuggling—be in perpetual crisis? In a diminished bureaucracy, the mid-level manager is part of a void in policymaking and divestment of resources that causes responses to common phenomena like unauthorized migration to be ad hoc, until the crisis becomes the norm. Similar narratives of crisis follow states offshore as they intercept and detain in ever more isolated corners.

What possibilities do territorially ambiguous zones offer for increasing understanding of sovereign practices? They are associated with distance from mainland sovereign territory. They seem at first external to sovereign territory, yet they connect intimately with practices and crises happening within sovereign territory. Agamben finds productive enforcement agendas hidden in these paradoxical binaries:

> It is almost as if, starting from a certain point, every decisive political event were double-sided: the spaces, the liberties, and the rights won by individuals in their conflicts with central powers always simultaneously prepared a tacit but increasing inscription of individuals' lives within the state order, thus offering a new and more dreadful foundation for the very sovereign power from which they want to liberate themselves. (Agamben 1998, 121)

The ambiguous sites between states connect with ambiguous legal zones and practices currently at work *within* states (or the sites conceived of more traditionally as sovereign spaces). They may appear stateless, yet are part of the state whether through devolution, contracting out of management, or investments in direct enforcement by civil servants. They are imbued with the state and render migrants stateless by geographical design.

Interstitial spaces tell us not only about shifting enforcement practices pertaining to immigrants and refugees, but also about nation-states themselves and about what happens when they allow for conditions of crisis. Drastic actions are taken; sites such as Guantánamo Bay, Esquimalt, and the dispersed and chaotic detention sites in Iraq emerge, threatening human rights and the integrity of UN conventions designed to protect them. The mere existence of stateless zones suggests an urgent need for renewed attention to the geography of nation-states, particularly to connect practices "at home" with those abroad.

Hiding people offshore mirrors the hiding of people onshore. Detention at Guantánamo Bay strips rights and legal protections from detainees, as the Patriot Act strips citizens and noncitizens of civil rights in the name of protection from ambiguous "others" elsewhere. Antiterrorism legislation in the regions discussed in this chapter has normalized crisis. Immigrants were quietly detained in county jails across the United States (Welch 2002). The conflation of discourse criminalizes nearly anyone on the move (terrorists, refugees, antiglobalization activists). In the UK, immigrants and refugees are those most likely to be detained under the provisions of the Anti-Terrorist Act. The same is true of antiterrorism legislation enacted in Australia and Canada following September 11. Among the victims of this violence are potential refugees and economic migrants.

The trends mapped in this chapter advance the argument that nations capitalize on crises in order to exclude. Whereas previous chapters demonstrated the anatomy of exclusion in Canada, this chapter has demonstrated that Canada's crisis-driven response to human smuggling resonated with enforcement strategies elsewhere, where states' being in crisis at home gives rise to statelessness offshore. Enforcement practices grow more transnational, and the relationships between migrants, refugees, and nation-states grow more ambiguous as nation-states manipulate geography to deny access.

Crisis combines with voids in policy and anti-immigrant discourse. Governments take advantage of crises to advance enforcement agendas. What may have appeared to be aberrations—a fleeting moment in Canada, a few people practicing interdiction in Hong Kong—proves to not be so when connected to the broader archipelago of enforcement practices occurring globally. The combination of antirefugee discourse with antiterrorist legislation and the extension of border enforcement beyond sovereign borders produce an environment in which spaces of protection for asylum seekers shrink daily. National exceptions become international norms, and transnational reverberations of threatening "others" elsewhere influence the daily lives of those racialized as "other" in the national imagination at home (e.g., Honig 1998; Puar 2007). Exclusionary state practices reverberate through the daily lives of migrants, immigrants, and citizens through mechanisms outlined in chapter 6.

Chapter 6   **In the Shadows of the State**

> Recently, I have been thinking about political subjects
> as subjects who imagine their position in the world and
> act, but who are also subject to state power. I have been
> thinking about the way the state manifests its power,
> especially when it does not call attention to its presence.
>
> —Wahneema Lubiano, *Mapping Multiculturalism*

THIS BOOK BEGAN inside the bureaucracy, but has moved gradually away from the office tower to the border and beyond, to extraterritorial sites where policing transpires.

This circular trajectory corresponds with my own research program that brought me into ever-closer encounters with the state. My work with undocumented Mexican migrants began with those who had recently arrived in the small city of Poughkeepsie, New York, where I grew up. In the early to mid-1990s I conducted a transnational ethnography, circulating between Poughkeepsie and a small town in Oaxaca, Mexico (Mountz and Wright 1996). While the state may appear to be absent from the lives of undocumented migrants, it is in fact a pervasive presence that structures where and how people without documents move through daily life. The U.S. government's interventions in undocumented migrations across the southern border wax and wane according to the economic and political environment of the time (see Nevins 2002). Early in his first administration, for example, President George W. Bush advocated for a guest worker program that would legalize temporary work migration from Mexico. After the terrorist attacks in 2001, however, he infused tremendous resources into enforcement along the border and internally, well within U.S. boundaries, to repatriate undocumented migrant workers.

147

In the 1990s I undertook a second transnational project on migration, working collaboratively with other geographers and grassroots organizations to research the residential and work experiences of Salvadoran asylum seekers with Temporary Protected Status (TPS) and to participate in their efforts to lobby the United States Congress for residency.[1] Despite extensive political organizing throughout the 1990s, Salvadorans continued to face a low acceptance rate for asylum in the United States. After experiencing many years of legal limbo, only 3.3 percent of Salvadoran applicants received asylum in the United States in 1997 (United States Committee for Refugees 1998b). Due to the lingering effects of cold war geopolitical relationships, these acceptance rates remain significantly lower than those for applicants from communist countries of origin, such as Cuba and Nicaragua.[2] I became involved in documenting Salvadorans' experiences of displacement and their efforts to lobby the U.S. government for amnesty. Asylum applicants spoke of the powerful ways in which the nation-state influenced their daily lives. As a result, I grew more interested in the uneven application of asylum policy on the part of the government of the United States. Eventually, what emerged from this work of gathering migrant narratives was a desire to also seek and add narratives of civil servants to the dialogue on immigration. This was an effort to understand, however imperfectly, the ways that the state saw and ordered transnational migrations.

As I documented the daily experiences and histories of Mexican and Salvadoran migrants in the United States, I came to understand that we inadvertently reproduce state policies in the intimacies of our daily lives. We enact the state. I grew more perplexed by the injustice and unevenness of policies regarding immigration and asylum and more interested in the state as an overlooked object of analysis in the field of immigration. I came to this research through a series of circular geographies connected to home and my own continuous search for home. As Elspeth Probyn (1996) writes, we are all "outside belonging," nomadic in search of a sense of place. That said, I write from a place of legal security and citizenship never to be taken for granted in a global environment where enforcement strategies increase people's vulnerability to statelessness (chapter 5) and the number of internally displaced persons receiving assistance from the UNHCR stands at 12.8 million, an all-time high (UNHCR 2007). Some people are able to become hypermobile citizens (Ong 1999) or, conversely, to simply stay in place, while others have less and less control over their own mobility (UNHCR 2007). It is essential to document this growing disparity by mapping transnational migration. This chapter engages the daily lives of those most affected by tactics of exclusion, those who live in the shadows of the state and yet know its power intimately.

Wahneema Lubiano suggests that "the state thinks the subject too" (1996, 65). But how does the state think the subject? Rather than look directly at the state apparatus here, I examine the effects of governance, the idea of the state and its reproduction and enactment on the ground.[3] This chapter addresses encounters with the state among people with varied legal relationships—U.S. citizens, residents with temporary status, and residents undocumented by the state—in order to map the technologies of power operating among them.

Body as Border

Raúl[4] is a New York–born U.S. citizen. In January 1999 he found himself in the Pacific Northwest, in the small city of Bellingham, Washington, about twenty-five miles south of the United States–Canada border. After spending the day job searching, he awaited the Greyhound bus traveling north along the coast to Vancouver, British Columbia. He intended to cross the border, the political determinant of the space where he is a citizen and the space where he is "foreign." Well dressed, he carried a new briefcase as he waited, vaguely aware of his surroundings, at the Bellingham bus station. As he sipped his coffee, he anticipated the usual troubles crossing the Canadian border as a Latino man. Suddenly three Caucasian, male plainclothes U.S. immigration officers approached him. After questioning his place of birth and travel destination, they identified themselves and requested a passport. Raúl, in an interview that followed, explained that he had long before learned to anticipate who would be questioned at borders.

Raúl: So I was drinking my coffee, and three white gentlemen came over to me and started asking me if I knew where that bus was going. And I answered them . . . and then actually that's when I realized that they looked like cops . . .

Alison: How did you know?

Raúl: Just because you grew up knowing their authority look, just the haircut, the jeans, you know what I mean? And then the way they were standing too, sort of like "at ease" with their backs against the wall. And then at that moment when I was thinking about it, he pulled out the badge and said he was from the border, immigration, U.S. immigration. So I put my coffee down, and I showed them my passport. And then they went, "New York, huh?" And I said, "Yeah." And they said, "Where in New York?" And I said,

"The Bronx." And they went, "Ah! The Bronx!" So they were talkin' how they knew cops from New York. And one guy said, "Fort Apache." And I said, "That's right, Fort Apache."

"Fort Apache" refers to one of the more abusive police stations in New York, against which members of the surrounding Bronx community, primarily working-class people of color, rioted in the 1960s. Raúl and the officers, through a brief exchange, had historicized their encounter. They referenced a relationship between state and other in another place and time. The officers read Raúl as a racialized body and, in spite of their initial attempts to mask authority, he located them too: their whiteness, their job. They evoked tropes of violence from his past.

Raúl: They looked at [the passport] and they said, "OK," and they walked away. But then I realized there was one of them sitting on the other side waiting for me to take the bus, and pretending he was a passenger.

Alison: How did you know he wasn't a passenger?

Raúl: Because first of all, the way he was dressed. And then when the bus was leaving, he stayed there sitting. He never got on the bus. And he kept staring at me.

Alison: And what about the bus driver?

Raúl: The bus driver, he was kind of funny. At first he went into the station, and when he came back, he was lookin' at me. So they must have said something to him when he went in like, "Keep an eye on him."

Alison: Now when you got stopped in Bellingham, it was surprising right?

Raúl: It's not the border. That's what shocked me. Why would they stop me where it's not the border?

The state is a powerful, dynamic entity rehearsed by a range of actors in everyday life. This chapter explores how people—especially migrants or U.S. citizens who are persons of color like Raúl and therefore racialized as foreign by border patrol—enact the state in their daily lives as the state attempts to simultaneously see and structure their mobility. According to Foucault, "[T]he individual is not a pre-given entity which is seized on by the exercise of power. The individual, with his identity and characteristics, is the product of a relation of power exercised over bodies, multiplicities, movements, desires, forces" (1980, 73–74). People experience the power of the nation-state distinctly in different places. The officers in Bellingham racialized Raúl, and he recognized their proclivity to do so. I, on the other hand, as a *Caucasian*, New York–born U.S. citizen sipping the same cup of coffee in the same place at the same time would probably not be approached with the same questions, or approached at all. Experiences of surveillance, though read off the flesh, do not leave unmarked the soul that inhabits the body. We are assigned guilt crossing borders, just as urban students passing through metal detectors are assigned guilt attending school, the violence of their environment scripted onto the body.

Nation-states control legality and police borders, criminalizing existence on one side or the other (Heyman and Smart 1999). Legally, Raúl's relation to the state is simple: he is a domestically born United States citizen. In many cases, however, as in this one, the United States government operates internally to police sovereign territory. Federal authorities police borders to protect nationals from foreign others. They script Raúl and position themselves by referencing the protection of a fort against racialized others, native or immigrant. In the country with the highest rate of incarceration and well-documented practices of racial profiling, they watch Raúl—a U.S citizen—to protect him from himself.

Of course, states see any migration imperfectly, and this extends to their efforts to categorize according to legal status, race, and location. Located elsewhere, along the U.S.–México border, for example, Raúl's body would not be categorized in the same way. Nations see in a variety of ways, not only through categories of race, but also with the assistance of biometric data gathered with new technologies and assembled through databases.

As nation-states "lose control" through economic and political deterritorialization (Glick-Schiller et al. 1992; Sassen 1996), the global economy dissolves borders for some. The nation-state, facing crises of sovereignty (Appadurai 1996, 169), reasserts sovereignty through the reterritorialization of other sites. Saskia Sassen identifies two such sites as body and border (1996, 65; 2006). I envision these two as one: border as body, body as the site where enforcement authorities encounter and reconstitute border. Raúl embodied and evoked the border some twenty-five miles from its physical location. As we interact, we socially construct, geographically imagine, and spatially reproduce borders. Raúl's was a relatively peaceful crossing, compared to the thousands that occur daily around the world, many of which are violent (e.g., Nevins and Aizeki 2008). And yet it is precisely the inevitable friction of one crossing compared to the ease of another that alarms.

Routine assaults provoke what Anzaldúa calls the "consciousness of duality" (1987, 37) in those made more visible by borders, which are evoked by skin color, dress, or accent. Such encounters with the state can criminalize and result, as in this narrative, in a potent form of self-surveillance. Raúl learned to predict who would be questioned, to anticipate trouble, and therefore to experience anxiety when approaching borders. Such apprehension feels familiar to all who approach borders, yet the degree of its intensity corresponds to the extent to which we know ourselves to be visible.

Not all border crossers internalize surveillance, however (see Heyman 1999), which underscores the point that populations are not always legible to states and that particular characteristics determine legibility and illegibility. While wealthy "globetrotters" experience international borders as increasingly porous sites of crossing, still others encounter them as increasingly militarized sites of immobility and surveillance. These contradictory experiences of the border spark a debate regarding the wellbeing of the nation-state in a transnational era. Is the nation-state a dying institution that has outlived its utility in a global economy (Appadurai 1996), or is it healthy, rearticulating for a global economy, operating flexibly, and creating varied forms of citizenship and mobility (Carter 1997; Ong 1999)? Viewing this debate through the lens of immigration policy and the geographies of its enactment provides some insight.

Our enactments and encounters with the state occur everywhere in daily life, well beyond the sites where borders are established. These encounters inscribe identities onto the body (Pratt in collaboration with the Philippine Women's Center 1998). Contrary to ruminations on the death of the nation-state, it is indeed alive and thriving, manipulating the porosity of borders differentially for distinct populations, and constructing vulnerable labor forces (Ong 1999) with tenuous or temporary legal status. Whereas I began with Raúl's experience, a moment of hypervisibility for the "disciplinary individual" (Foucault 1995, 227), I will move beyond internationally constructed borders to analyze the operation of disciplinary power that renders bodies visible in other spaces. People with varied mobility and legal relationships to the United States government enact the state, providing testimony to the power of the nation-state to influence the intimacies of everyday life.[5] As political debates about immigration policy and guest worker programs escalate in the United States, daily enactments of the state illustrate that it is not merely an institution or set of power relations "out there," but a set of practices, processes, and expectations that we operationalize very much "in here."

Transnational Panopticism

I offer here transnational panopticism as a tool to conceptualize the relationship between state and individual by locating the exercise of sovereign power at the scale of the body. The same networks of mobility and communication that constitute transnational space (Glick-Schiller et al. 1992) also serve as networks of surveillance. Vision, knowledge, and the operation of power form the core of this concept. Also central is the relationship between body and border. In the geographical testing of the conflicting narratives of the state, the body emerges as a key site at which to examine the performance of sovereignty. The state performs visibility distinctly, operating differentially on distinctly scripted bodies in distinctly coded spaces. Immigrants also construct identities discursively, making themselves more or less visible in different contexts. As Michael Taussig (1997) suggests, "[I]t's not so much the rule that has been broken but a matter of knowing the rule or even knowing that there is one" (19).

Glick-Schiller, Basch, and Blanc-Szanton (1992) provide a seminal and enduring definition of transnationalism as "the emergence of a social process in which migrants establish social fields that cross geographic, cultural, and political borders" (ix). While transnationalism is not a new phenomenon, the *pace* of connections has altered, corresponding with the late stages of capitalism and people's attempts to negotiate the global economy with multiple identities (Kearney 1991; Glick-Schiller et al. 1992; Rouse 1995; Mitchell 1997a; Ong 1999). As Roger Rouse remarked, "[T]he raw materials for a new cartography ought to be . . . discoverable in the details of people's lives" (1991, 9).

Indeed, in the alluring theoretical world of time–space compression (Harvey 1990), globalization, and free trade free-for-alls, scholars have been drawn to the conceptualization of explosive practices of mobility that prompt new spatial arrangements (among them, I contend, arrangements of the state). Transnationalism signals flows of capital, information, and people, and consequently, we write about the ease of movement and the disappearance of borders (for critiques see Mitchell 1997a; Cunningham and Heyman 2004). Much discussion of globalization, for example, occurs over economic terrain where free trade dissolves borders and nations lose sovereignty. And yet the status of the nation-state remains ambiguous and contentious. Some look beyond an "obsolete" nation-state (Appadurai 1996, 169) for potential in a "postnational" world (Appadurai 1996; Anderson 2000; Soysal 2000). Whether peering into quotidian details of transnational life or measuring flows of transnational capital, our attention has been diverted from the role of the state, the effects of policy, and the transnational realities of immobility and exclusion.

Yet state policies reproduce difference and disparity through immigration policy. One example is the creation of a vast, "unauthorized," vulnerable, temporary labor force in the United States, estimated to number some

12 million. Such articulations, however, are not always discernible from readings of policy alone; as Lubiano remarks, "One job of the state is to make its manifestations of power in the operation of the economy culturally acceptable" (1996, 71). A look at the effects of immigration policies reveals the extent to which they strip immigrants of their rights to organize and to access social support, education, and health resources.

Transnationalism arose as a theoretical framework designed to challenge conventional assumptions of state-centric immigration models that quietly followed the temporal and spatial logic of assimilation (Kearney 1991; Glick-Schiller et al. 1992; Rouse 1992). Early theorists of transnational migration critiqued structural immigration models for viewing migrants as pawns being pushed and pulled around the world according to fixed economic contexts and temporal trajectories. Glick-Schiller, Basch, and Blanc-Szanton (1992, ix) instead thought of transnational migrants as an aspect of the complex decisionmaking central to global processes. The literature is replete with calls for "transnational spatial ethnographies" (Mitchell 1997b, 110), "global ethnoscapes" (Appadurai 1996, 64), and an "anthropology of transnationality" (Ong 1999, 240). With much attention devoted to migrant subjectivity, however, this work was in turn critiqued for failing to understand the structural contexts influencing migrants' decisions, such as material impacts of globalization and development (Silvey and Lawson 1999, 123). Now scholars seek recursive theories that mediate fluidly between human agency and structural factors without simplistically dividing the complexity of life between the two.[6]

A Foucauldian analysis offers one possibility for conceptualizing the operation of power within transnational populations. Foucault used Bentham's Panopticon to exemplify his conceptualization of disciplinary power (1995, 171). The circular design of Bentham's prison houses cells along the perimeter, surrounding a central watchtower occupied by an anonymous individual. Key in the relationship between the guards of the central tower and the prisoners in cells is visibility: the occupant can see any prisoner at any time. While the prisoners cannot all be seen simultaneously, they nonetheless know they can each be seen at any moment. They therefore practice self-surveillance and self-regulation. Bentham's prison compartmentalizes space and infuses it with meaning and subjectivities. A self-regulating "efficient machine" from which no one can escape, the Panopticon specifies a particular architectural design for surveillance. Theorists have employed this design to interpret landscapes of fear and control, among them the geography of immigration policy as a form of transnational panopticism.

Fundamental to the operation of disciplinary power is knowledge of the body. "The gaze is alert everywhere" (Foucault 1995, 195), making every action immediately knowable (1990, 173). This knowledge of observation mobilizes individuals like Raúl to practice self-surveillance. This discipline operates in distinct ways, on distinct bodies, in distinct places (Foucault 1980, 158). Foucauldian space is coded, classified, partitioned, and endowed with meaning. States demarcate and regulate space (1990, 195) with immigration policy. Although migrants utilize networks to cross national borders, panopticism—the possibility of being seen—encourages them to practice self-regulation more effectively than the nation-state can monitor them. Although authority may appear absent at any given time or place, the perpetual assumption that we are always watched remains, particularly for those undocumented by the state.

As Foucault argued, "[E]xcessive insistence on [the state's] playing an exclusive role leads to the risk of overlooking all the mechanisms and effects of power which don't pass directly via the State apparatus, yet often sustain the State more effectively than its own institutions, enlarging and maximizing its effectiveness" (1980, 72–73). Disciplinary power is productive rather than a repressive force, productive of identities in relation to immigration policy in daily life (e.g., Pratt 1998). Like borders, immigration policies are devised, mobilized, and articulated differently for different people. Thus, through the artful practices of government policy, belonging to the nation-state becomes more robust for some and more tenuous for others, which means a more secure life for some, supported by the more perilous lives and jobs of others. The idea of *governmentality* (Foucault 1991) entails the functioning of the nation-state more powerfully in geographically dispersed zones of the social body, beyond the spaces of government institutions. Tenuous legal status—whether temporary or undocumented—relegates certain workers to certain locations and niches in the labor market, while entitling others to a sense of belonging to a state accompanied by socioeconomic senses of place.

Spatial technologies of visibility, knowledge of the body, and self-surveillance emerged clearly in Raúl's border crossing. His ability to predict which passengers on the bus would face questioning illustrates the operation of disciplinary power on distinctly racialized bodies. As "disciplinary subject," routine and racialized interrogations incur feelings of anxiety and guilt. They have trained Raúl to know he is being watched and to exercise self-surveillance when approaching borders, internalizing and sustaining the power of the gaze while also recognizing and resisting it. Regardless of legal status, Raúl was deviant.

The Department of Homeland Security (previously the Immigration and Naturalization Service) deliberately maintains a foreboding presence in sites where territory is strictly demarcated and regulated. American geographical imaginations construct borders as powerfully fixed gates and fences, policed by ever larger and better-armed troops of INS officers. Reports on Operation Gatekeeper and other projects to strengthen U.S. borders with the use of new technologies of surveillance appear frequently in the media. To enter a DHS building, one must pass through multiple checkpoints and metal detectors.[7]

The architectural effects of the prison on prisoners and the border on noncitizens parallel those of immigration policies on immigrants. Contrary to popular images conveyed by the media, borders are enacted not as fixed lines but as ubiquitous, liminal zones. Imagine a transnational model of Bentham's Panopticon wherein those same networks of communication, exchange, and mobility through which transnational space is produced simultaneously function as nodes of surveillance. Borders are alive and operational wherever we reproduce them in daily life.

Transnational panopticism serves as a fluid conceptual tool with which to examine relationships between states and individuals. It operates distinctly where individuals have different relationships with the state. Transnationalism by its very definition implies complex and far-reaching networks that link individuals and communities located in multiple places. Likewise, the enactments of the nation-state occur transnationally. These relationships straddle scales, including the transnational individual, family, and community, each asserting the possibility that "the local" and "the global" are separate yet intimately intertwined. The state is enacted in daily discourse, weaving its way in and out of ethnographic accounts of mobility. Migrant narratives reproduce the state where it might otherwise appear absent (Lubiano 1996).

Meanwhile, as policies render individuals visible and vulnerable, they remain confusing and inaccessible in their uneven design and enactment.[8] Examination of the daily manifestation of policies reveals the ways in which individuals grapple with issues of belonging and exclusion. Kearney's "border ethnography" eloquently illustrates the process of criminalization of migrants crossing the Mexico–U.S. border (1991). His argument can be extended into internal sovereign territory, where borders are constructed everywhere in transnational daily lives—where bodies are borders. This interior policing has become more pronounced in recent years with 287 arrangements between federal and local authorities to share responsibility in policing undocumented migrants (Coleman 2009).

Transnational panopticism links migrants to the border in daily ways through technologies of seeing. In Oaxacan villages, residents watch for remittances to determine success or failure, the fruits of their neighbors' labor in the United States: "He is working. . . . This family . . . well, you can see what they are doing. You can see that they are constructing" (Mountz 1995, 170). This watchfulness extended beyond economic investments to the policing of gendered social relationships. One young Oaxacan woman said, "Although it seems like you alone are doing your things? I don't know—there is someone who is watching you. And they put this someone in charge of publishing it" (Mountz 1995, 176). Women such as Luisa, who were among the first women to migrate from the small Oaxacan village, were punished through the steady stream of transnational gossip, deemed "loose women" for having transgressed social norms through their border crossing.

Panopticism thus emerges in everyday ways within transnational communities on both sides of the border and holds potential for uncovering material manifestations of immigration policy in daily life. For some the state is highly visible, and for others the state recedes, though its power to act remains. Whereas Raúl's disciplinary experience remained within close proximity of the state apparatus, disciplinary power operates well beyond the border.

I turn now to two other cases of transnational panopticism involving contemporary migrations from Central America to the United States. I examine the confinement of Salvadoran asylum applicants—a group for whom the state looms large—granted temporary status in the United States, and a community of undocumented Oaxacan migrants for whom the mere *idea* of the state shapes daily life, for, according to Foucault, "what matters most is that he knows himself to be observed" (1995, 201). Transnational identities are locally gendered, geopolitically bound, and globally produced.

Strategic Visibility

The state creates vulnerable labor forces by constructing ambiguous forms of belonging through immigration policy. The increasingly popular practice of granting temporary or guest worker status constructs immigrant populations as "temporary" and thereby denies them the rights and resources accorded "legitimate" residents of the nation-state through citizenship.

Michael Kearney points out the practice within the United States of bringing migrants as workers, but not as citizens or residents with rights and
resources:

> The individual migrant is allowed into the unitedstatesian nation-state not as
> a citizen, but as an 'alien,' not as someone to be incorporated into the social
> body, but as someone to be devoured by it. Migration policy/policing and
> resistance to it is thus a struggle for the value contained within the personal
> and social body of the migrant. (1991, 63)

Many Salvadorans living in the United States today find themselves
in this situation. Most immigrated during the Salvadoran civil war from
1979 to 1992, entered the country unauthorized by INS, and either
applied for asylum or remained undocumented. Throughout the 1980s,
while the United States continued to support the Salvadoran military with
over $1 million per day, asylum acceptance rates for Salvadorans were
approximately 2 percent, compared to about 26 percent for Nicaraguan
applicants. This practice was challenged in the mid-1980s in a class action
lawsuit filed by the American Baptist Church against Attorney General
Thornburgh. The suit was eventually settled out of court and resulted in
the creation of the ABC program, which coincided with the beginning of
Temporary Protected Status (TPS), granted to Salvadorans in 1990. Thus
began a series of temporary programs that would expire and be renewed and
renamed throughout the 1990s.[9] In September 1998, 181,090 Salvadorans
had pending asylum applications (United States Committee for Refugees
1998a, 6). As of November 2004, the number of Salvadorans with TPS
had grown to approximately 290,000, by far the largest nationality group
with TPS (Wasem and Ester 2004, 6). The programs provide asylum applicants permission to reside and work on a temporary basis, pending annual
reapplication. Without the promise of permanent residency, this law makes
people appear to lead temporary lives. And yet many resided in the United
States for ten to twenty years of temporary status without being granted
residency. Asylum acceptance rates among Salvadorans remained at about
4.7 percent throughout the 1990s (United States Committee for Refugees
1998a, 6–7). The effect—however unintended—of these programs was to
maintain displaced immigrants in legal limbo.

Since the 1980s many advocacy organizations and immigrants with
temporary status participated in a grassroots political movement to acquire
residency. They pressured the U.S. government to recognize Salvadorans as
refugees and as immigrants (Coutin 1998, 906) and to grant them asylum

and residency. A politicized battle raged from the late 1980s through the 1990s over the social construction of Salvadorans as "deserving refugees" or "undeserving illegal aliens" (Coutin 1998, 906). However, not much changed for the vast population of "temporary" immigrants in search of permanent status in the United States.

Throughout the conflict in El Salvador, both the leftist guerrillas and the U.S.-trained military waged war on Salvadorans. Anonymity was impossible as they used extreme tactics of recruitment, repression, and torture, imprinting the relationships between identity and surveillance with violent markings on the body. Uncertain status led to practices of self-surveillance and self-regulation. Because of their history, Salvadoran asylum applicants have a heightened awareness of the relationship between visibility, knowledge, and power, and they equate surveillance tactics with U.S. involvement in the Salvadoran conflict. Many suffered from post-traumatic stress disorder (PTSD), fearing routine interactions with such individuals as police, hospital and clinic workers, and immigration officers. One man reported a fear of tollbooth operators, whom he believed monitored his movement. Many live in low-income households yet pay monthly subscription fees for caller ID to render callers visible and themselves less vulnerable. The temporary status programs exacerbated PTSD by prolonging uncertainty about home and belonging in formal and informal spheres of citizenship. The feeling of being watched was pervasive, perpetuated by the experience of a criminalized existence in the United States and the psychological trauma caused by insecure trajectories.

Fear and insecurity are reproduced by the state on a regular basis.[10] Applicants must reapply and pay for their work visas annually. Temporary visas ensure instability in the labor market, where the stigma attached to legal status results in employer abuses that parallel those of undocumented workers (Wright et al. 2000). Residency remains a distant hope and challenges decisions to invest in the future. Temporary policies render applicants visible by requiring immediate notification of a change in address, telephone number, or place of employment. Furthermore, applicants may not leave the United States for any reason, nor do they have rights to family reunification sponsorship. Many parents who left behind young children had not seen them in over ten years, unable to return to El Salvador or to bring them to the United States (Miyares et al. 2003). Salvadoran transnationalism for those in temporary programs was thus largely characterized by absence, immobility, and the inability to seek closure on the past and progress into the future (Bailey et al. 2002). Conditions of participation in temporary status programs therefore entail potent forms of social control (Driver 1985, 427) with visibility, spatial entrapment, and the "art of distributions." These are techniques that involve enclosing, partitioning, and coding space and establishing rank (Foucault 1995, 140–45). Temporary status is one among a hierarchy of legal categories that function as artificial practices of classification to control the rights and mobility of individuals. Facing an unfortunate and maligned convergence of historical factors, residual cold war politics, and geopolitical amnesia, those with Temporary Protected Status were among the most regulated of immigrant communities in the United States.

Temporary status programs serve as exemplary models of disciplinary power imposed by the state, shaping daily decisions with little direct exertion by the state. A disciplinary discourse has developed around temporary status programs and the conditions of asylum approval or denial. Through their interaction, asylum applicants, community advocacy organizations, and the INS construct notions of "strong cases," ideal behavior, and hence "good" Salvadoran immigrants. In order to support temporary status applicants through the complex legal proceedings, advocacy organizations regularly give talks and distribute materials to explain what the temporary programs mean and how to negotiate them. Coutin explains the tendency on the part of legal advocates to measure asylum cases for their potential (1998, 914). Asylum cases are determined not only on the basis of history of persecution in El Salvador, but also on the basis of participation in the U.S. economy and society. This corresponds with the shift on the part of Salvadorans and legal advocates to change the designation of Salvadorans from "refugees," with certain rights, to "immigrants," with other rights (Coutin 1998), from persecution for which the United States held partial responsibility to settled, participating, taxpaying residents to whom the United States holds other responsibilities.

Behavior constructed as "good," or as promising a potentially "good American," illustrates for INS judges a settled life that contributes to U.S. society (Kerner et al. 2001). This idealized individual partakes in heterosexual marriage, steady work, home ownership, payment of income taxes, organized civic life, and English classes. He has U.S.-born children in school, bank accounts, and investments—precisely the type of settled life inhibited by the conditions of the temporary status programs. Evidence of this discourse emerged in our data statistically, with spates of marriage among partners who had cohabitated in civil unions for long periods of time, and in political organizing through which collective identities were constructed with protestors' displays of tax forms.

People with temporary status are required to practice and document this behavior. Legal advocates encourage letters from church ministers and from English teachers confirming attendance. This ideal citizen stands in opposition to the immigrant constructed in public discourse as a non-contributing drain on the system, "taxing" U.S. society and government by using medical services, welfare, food stamps, unemployment payments, and benefits at taxpayers' expense (Coutin 1998).

Salvadorans interviewed did not participate in certain activities, regardless of whether or not they were regulated directly by state authorities, for fear that they would destroy their identities as potentially "good" citizens, and therefore "strong" asylum cases. Many feared partaking in labor unions and avoided utilizing social services granted to "participating" U.S. residents. The following quotes from applicants residing in the United States for some ten years illustrate negotiation of these identities.

Julio: There are more people who do not want us here than there are those who do. . . . And what hurts is that one comes here to work. I contribute to this economy. I pay taxes, rent, phone, light, the same as Americans. I am American. They discriminate against us for a million reasons. . . . Now they want to kick us out after what they've done. After doing one damage, they want to do one that's worse. . . . I have now learned to live in this society. . . . It would be hard for me to return to El Salvador now. We are not asking for much. I'm not going to live off the government. I live off of my work.

Marta: I make up part of the society, but they don't accept me as such. . . . We are discriminated against because we are a minority and because we came from underdeveloped countries. We are not, according to them, contributing to the needs of the country. . . . We were fleeing because of a war that the very United States generated. . . . So one has to leave the country, risking her life. . . . If they didn't kill you in the war, they kill you in the street when you come to the U.S. And if you succeed in passing those two barriers, you won. But if inside here, they don't give you papers, then you are the same. You continue as a delinquent fleeing. Fleeing something that you have not done.

Homeland Security (formerly the INS) perpetuates histories of surveillance by monitoring temporary status applicants, and immigrants internalize and maintain this fear through self-surveillance. Temporary status programs have the unintended consequence of maintaining an exploitable workforce of individuals afraid to protest abuses such as poor conditions and frequent dismissal.

In their political struggles, immigrants and advocates collectively construct visibility of a particular sort. Community groups carefully organize speakers at protests in public spaces and make signs to illuminate certain behaviors such as paying taxes or working two jobs. They strategically stage children and long-term residents to speak. Some protest with signs that portray enlarged W-2 forms to demonstrate payment of taxes. This publicity stands in striking contrast to other sites where Salvadorans demonstrate invisibility, such as in utilizing social services.

In 2006 a series of public demonstrations in support of more permissive policies toward undocumented workers in the United States signaled the reopening of political debates over amnesty and naturalization. Children again marched with signs that said, "My mom has two jobs." The process exemplified Foucault's "economy of visibility" (1995), objectifying and fixing the subject; documenting histories as a means of control and domination (1995, 189–191); and culminating in the asylum interview, where applicants attempt to prove events difficult to document, such as persecution, torture, and evidence that a return to El Salvador today would cause "extreme hardship." After significant lifestyle adjustments, identity maneuvers, legal fees, anxiety, and years of legal and personal limbo, applicants still faced a 4.7 percent chance of asylum. If denied, they must decide whether to appeal and continue legitimately through the process or to disappear underground by changing residence and place of employment, slipping into undocumented legal status. Through the interpretation and negotiation of asylum discourse, narratives reveal how people internalize state expectations but also resist self-regulation.

Strategic Invisibility

The state is powerful although it may in fact appear absent or disembodied (Taussig 1997). By definition, the state is supposedly not aware of the presence of undocumented or unauthorized migrants. This population offers the most compelling example of transnational panopticism and the power of the state, for despite its apparent absence from daily life, the idea of the state continues to shape intimate details. The daily lives of undocumented residents are highly influenced by the threat of detention and deportation, including where and how one lives, socializes, and works (Kearney 1991; Rouse 1992). Here transnational panopticism explains the disciplinary power of the state as experienced in daily life through the practice of self-regulation. Whereas Salvadoran asylum applicants construct collective identities to protest federal policies, undocumented migrants construct invisibility and live daily with a fear of being seen.[11]

One primarily undocumented community of migrants circulated for several years between two physically distant places: San Agustín, a rural village in the southern Mexican state of Oaxaca, and Poughkeepsie, a small city on the Hudson River, outside New York City.[12] Over time, the residents of San Agustín and Poughkeepsie forged one cross-border community across transnational time and space (Mountz and Wright 1996). Inhabitants of this community operated daily with a constant awareness of events occurring throughout the transnational community. One mechanism of transnationalism was a panoptic gaze rendering all residents visible through gossip. Residents used the same technologies of visibility that produced transnational economic and social relations as networks of surveillance. Despite a dearth of telephones and electricity in San Agustín in the early 1990s, Oaxacan residents in Poughkeepsie often knew about events occurring in the space of the village before the villagers did themselves.[13] Family members in Poughkeepsie would send televisions, VCRs, video camcorders, faxes, and eventually e-mail to San Agustín.[14] They sponsored celebrations, and their families, in return, videotaped the festivals and sent the tapes to Poughkeepsie. During festivals, transnational relationships changed as boyfriends abroad observed dance and conversation partners of loved ones at home. The public space of the village now includes the space of Poughkeepsie, and an active chain of gossip, or *chisme*, regulates the activities of all who move through this space.

In the early 1990s the vast majority of Oaxacan residents in Poughkeepsie lived an undocumented existence, having crossed the border without the sanction of the government. Unlike those with temporary status, residency, or citizenship, the state has no record of the presence of undocumented residents, and yet they lead daily lives shaped by constant knowledge of surveillance. Their undocumented existence was criminalized and regulated. The gaze, perpetuated by transnational residents, reinforces the power of the state to deport. Like undocumented workers across the United States, Oaxacans worked as an inexpensive labor force in Poughkeepsie with few organizing rights. They worked twelve- to fourteen-hour days for less than minimum wage. Employers abused workers by threatening to report them to authorities should they not comply with work conditions. When this does not occur explicitly, the mere possibility of deportation silences workers. For undocumented residents, the Panopticon operates as an efficient machine with the least amount of exertion on the part of the state apparatus (Foucault 1995).

Most undocumented Oaxacans moved around Poughkeepsie by bicycle or on foot and shared apartments. When attacked or robbed, they often feared reporting incidents to the police. In a dramatic expression of self-surveillance, migrant themselves occasionally threatened to report one another to authorities. It is in the daily existence most detached from the U.S. nation-state that we see the most organized practices of disciplinary power through extensive transnational technologies of visibility. In interviews, undocumented residents spoke often of being watched. Their narratives reveal ways in which they internalize the gaze, resulting in self-regulation parallel to that experienced by other border crossers.

These experiences are gendered. Women abused at home fear seeking the help of police and women's shelters because of the threat that police, social service agencies, or even their partners will document their presence. In many ways, the state is reproduced in daily life.

From Enactments to Encounters

Surveillance along the U.S.–Canada and U.S.–Mexico borders intensified dramatically after 9/11 (Andreas 2003; Gilbert 2007; Nevins and Aizeki 2008), the United States has still not granted residency to Salvadorans with temporary status, and an estimated 12 million people continue to live and work without recognition from the U.S. government. Guest worker populations, sometimes undocumented and other times documented in temporary form, are increasing in Europe and North America. For the undocumented, borders follow bodies, criminalizing and disciplining not only at the site of a person's crossing but everywhere in daily life.

Popular media, politicians, and state policy work in various divisive ways to set groups with differential legal status against each other. In 1995 Proposition 187, for example, threatened to strip undocumented immigrants in California of access to health and educational services and create a repressive atmosphere of surveillance that exacerbated animosity among groups differentiated by race/ethnicity and legal status. Citizens were required to report the undocumented. The proposition was, however, found to be unconstitutional, as it would have denied these immigrants and their children basic health and educational services. Several years later, the seeds of this proposition blossomed in the form of federal and local collaborative policing practices. Even where formal policies such as 287g did not exist, local civilian residents took up the work of border policing by lodging complaints about their neighbors.

Following the terrorist attacks of September 11, the Patriot Act reduced civil liberties and encouraged citizens to "keep watch" for criminal behavior, illustrating the continually imperfect enforcement tools of the state and its need to rely on others to govern immigration. Such rhetoric divides people and persuades them that there is an enemy against whom they live and struggle. Media, policy, and scholarship circulate tropes of immigration as invasion, flood, and balkanization (Lubiano 1996; Ellis and Wright 1998). This rhetoric distracts our attention from the reality that we are all policed more violently everywhere. In situations that feel comfortable legally, socially, and economically, we quickly forget our own family histories of mobility and displacement upon arrival.

As the scenarios in this chapter demonstrate, the nation-state is not simply an entity that exercises hegemonic power. It is also a powerful discursive construct. Raúl's knowledge of authority facilitates his ability to see through and around questions posed in Bellingham. Migrants crossing the border without legal sanction know the best times and places to cross. Salvadorans crossing into the United States from Mexico know to label themselves Mexican when apprehended, so that they will only be deported to Mexico.

Like authorities, individuals perform visibility in particular ways to advance political agendas. Some Salvadorans construct public identities that challenge policy while others resist by slipping away from temporary status programs into an undocumented status. Women share information and resources to support each other's migration to Poughkeepsie. Within Poughkeepsie, immigrants share knowledge about which service agencies do not require documentation. Resistance takes the form of networks of shared knowledge. Immigrants construct identities that simultaneously strengthen and subvert the idea of the nation-state and its wishes for migrants.

The state is enacted in embodied, everyday ways, the body being a site where global power struggles are negotiated and identities constructed. After reading the migrant through the state and human smuggling through the eyes of civil servants, I end here by reading the state through the body, its panoptic gaze inculcating particular practices through feelings of fear and insecurity in daily life. Just as states encounter, read, and reproduce individuals, so do individuals reproduce the nation-state.

There are symbiotic links between the dilemmas and possible courses of action confronting migrants and civil servants who negotiate border crossing and other moments in the process of transnational migration.

Time is key, whether it be the drawn-out wait for asylum or crossing, or the moment of crisis. The state advances enforcement agendas that are hyperracialized and performative when security agendas come to the fore. Furthermore, civil servants and migrants alike enact and embody the policies of the state unevenly, and struggle with the knowledge of this uneven enactment.

Chapter 7   What Kind of State Are We In?

> If we pause for a moment on the meaning of "states" as the
> "conditions in which we find ourselves," then it seems we
> reference the moment of writing itself or perhaps even a
> certain condition of being upset, out of sorts: what kind of
> state are we in when we start to think about the state?
>
> —Judith Butler (2007, 3)

THIS BOOK HAS EXPLORED the relationship between discourse and practice, between the production of mobile subjectivities—the smuggled, the refugee, the spontaneous arrival, the detainee—and their abjection. Contemporary discourse on migration and asylum is indeed riddled with metaphors of exclusion by states. Australia *excises* islands, Europe *externalizes* processing, and Canada crafts the *long tunnel*. These metaphors represent exclusionary geographies that contribute to the shrinking of spaces of asylum. Meanings of asylum, initially designed to protect, are themselves crossing into a new phase of securitization. The very border enforcement regimes developed to curb human smuggling also stop asylum seekers from reaching sovereign territory. The securitization of migration renders those persons in search of protection more vulnerable.

States themselves act as the architects of statelessness, by utilizing legal ambiguity, temporary policies, and detention centers. It is imperative, therefore, that we turn our attention not only to those sites and persons that are stateless by geographical design, but also to states themselves, to document and understand their practices. While civil servants work furiously to manage human migration, social scientists must work equally hard to trace the changing nature of sovereignty and the many contradictions involved in its exercises in border enforcement. We must write our way to and through the state, to understand just what kind of state we are in.

The book has examined state practices and their effects, from the conventional borders of sovereignty, to the preemptive transnational state that stops migration abroad. The analysis began at the administrative center and then moved progressively outward—along state borders and abroad to the international staging areas of human smuggling, to the margins of sovereign territory where asylum seekers find themselves detained, and then into daily life, where we negotiate and confront immigration and refugee policies. States enact violence not only in the act of detention, but also in the more mundane practices of exclusion, where migrants and asylum seekers are haunted as policy reverberates through work and life.

I have argued that states of migration are capitalizing on crises to advance enforcement agendas that exclude those in search of refuge. Whether viewed from the outside in or the inside out, the state is imagined, enacted, and encountered in our daily lives. In these processes, state borders expand far beyond political boundaries, moving outward to bring into being paradoxical zones and extraterritorial locales of policing and detention. The text dwelled in zones of exclusion: the tunnel, the detention center, the hearing, the island. These cartographies of enforcement demonstrate the securitization of borders that corresponds with shifting, privatized, dispersed, transnational areas of sovereignty. New political geographies of the state have been drawn, often constructing intimate and proximate boundaries around the body.

The competing views of borders held by civil servants and transnational migrants point to contradictory "states," where states act more transnationally in their enforcement practices, exercising sovereign powers that extend beyond traditionally conceived boundaries of sovereign territory. The book's ethnography of CIC has shown that, in the realm of human smuggling by sea, crises in the media often intersect with voids in policy, with exclusionary results. A battle in the naming and categorizing of those intercepted—typically, "bogus" refugees—accompanies remote institutional geographies of processing and detention. Canada does not act alone in these practices; rather, it is accompanied by other countries that are "leaders" in realms of border enforcement and refugee resettlement. States prove performative in their responses to human smuggling crises in the media, working their way productively—if perilously—through competing narratives of vulnerability and might with corresponding geographies of stateless spaces, remote detention, and interdiction.

Most migrants who came by sea to Canada were repatriated in highly publicized and controversial chartered flights in 2000. As for Canada, after years of negotiations to draft and sign an interdepartmental Memorandum of Agreement (MOA) that would fill the void in policy around interceptions, the MOA was finally completed but never signed, leaving Canada in the same state of unpreparedness. Whether written into policy (as in Australia) or kept geographically apart from access to policy (as in the United States), ad hoc arrangements intersect with national security imperatives, resulting in crisis and exclusion.

Localized, seemingly aberrant phenomena connect with broader enforcement trends happening across international borders. In a 2007 editorial called "Gitmos across America," the *New York Times* recognized the relationship between the detention of foreign nationals within and outside sovereign territory. Hidden and legally ambiguous detentions reduce chances of protection for those on the move. Exceptionalism prevails both on- and offshore.

Just as Gibson-Graham (1996) provoked readers to think outside the paradigm of capitalism, we must think beyond the inevitability of expansive borders, shrinking spaces of asylum, and "Gitmos across America"—in short, beyond the pervasive, mythical power of the magical state—to imagine and struggle for alternative geographies that protect and include, rather than endanger and exclude. Transnational migrations have elicited transnational enforcement responses, which in turn have elicited transnational activist projects. As states design sites that evade legal protections, a dynamic array of advocates and activists also work across borders to design alternative geographies of mobility. They confront transnational enforcement practices by drawing creative, transnational politics of resistance (Grewal and Kaplan 1994; Portes et al. 1997). National and transnational social movements are gaining momentum as people become politicized to oppose enforcement regimes, offering narratives to counter those of detention and exclusion. Their strategies render visible those practices and people that have been concealed by states. Australian activists wrote letters to asylum seekers detained on islands offshore without telephones or access to legal counsel (Lonely Planet 2003). They pushed down fences around the Woomera detention center in the Australian outback. European and American organizations such as the Detention Watch Network map detention to bring into view the spatial isolation where governments attempt to hide through detention. Activists in Texas and South Austrailia succeeded in shutting down detention centers. In Canada, Australia, and the United

States, detainees themselves resist with hunger strikes, escapes, and, in their darkest hours, suicide. In January 2009, 700 migrants detained on Lampedusa pushed down the gate to protest the conditions of the detention center (BBC 2009).

Mapping and Making
Ethnography of the State

Anthropologist David Valentine (2007) "makes ethnography" of a category; in his study, the category is "transgender." Valentine (231–33) writes:

> Like my trusty bicycle, on these nights transgender is a useful way of getting around, of going from one thing to another, of framing a set of diverse moments and social practices in time and space as an entity. At other times, though, this feeling dissolves into a new story, a fractured sense that I am the starring lead in an ethnographic unity of my own making . . . "the transgender community" . . . even as it exists and is real—is at the same time a product of an imagined unity.

My own project takes cues from Valentine. In parallel fashion, I have pursued the transient, troubled category of "the state," a category that is often more coherent in the writing of academics and activists than in the daily work lives of bureaucrats. The ethnography of the state that I have been "making" here has mapped states not only as institutional arrangements and geographical entities where power circulates through migration-related decisionmaking, but also as an iterative process. As such, an ethnography of the state cannot dwell in any one locale, but must move alongside the state itself. The sites in this book include the edges of sovereign territory and bureaucratic work where migrants, media workers, activists, lawyers, and civil servants struggle over states of migration. Transnational ethnography enables the material mapping of those sites that Agamben (1998) so intensely describes and yet overlooks as everyday practice. Only through ethnography do figures of smuggled migrants and civil servants become embodied and differentiated, rather than racialized and homogenized.

What alternatives to the securitization of migration can be found? State institutions work their way into the intimacies of our daily lives, but they are not without vulnerabilities and fissures of their own. These fissures are small cracks, but ripe with political potential. They become more visible in daily bureaucratic exchanges made up of individuals, groups, and networks across which information flows and policies are formulated.

Meanwhile, people and activities on the margins of the state illuminate its shifting center from the topography of its margins. These shifts are more visible in the acts of daily practice than in the texts of policy, but they require innovative methodologies and creative mappings.

Giorgio Agamben sheds light on the margins of the state, writing stories without mapping them. However, contemporary camps and state practices that render "others" stateless by geographical design must be understood on the ground—how they come into being and function on a day-to-day basis. Only by following the lead of anthropologists moving inside the inner workings of the state (Hansen and Stepputat 2001) can we come to understand *how* Agamben's states of exception emerge.

The daily negotiations occurring between state and migrant within and beyond sovereign territory must be documented. The theorizing of state activity in extraterritorial locales requires, in turn, a rethinking of transnational geographies of sovereignty and subjectivity (Hansen and Stepputat 2005).

One of the critiques of ethnography is its tendency to dwell in the details of life in one specific locale. This ethnography of the state, for example, hovered in particular in the office of employees of Citizenship and Immigration Canada as they struggled with human smuggling from China to North America. The ethnography responds to critiques of "smallness" in three ways. First, only by entering the anatomy of crisis, talking with and observing those working at its center, do we come to understand the precise moment when crisis gives way to securitization—when mundane border enforcement gives way to more aggressive exclusion of asylum claimants. Second, the connection of the Canadian case to the international context in which civil servants devised and learned strategies from other states sheds light on the fact that those who created the long-tunnel thesis did not act in isolation, but in the company of and conversation with others. Third, theoretical interventions by scholars working in the arena of enforcement and transnational migration facilitate connections between local cases and global trends. Ethnographic research offers the potential not only to record the long tunnel, but also to understand how it is a tunnel that opens endlessly onto others.

Travesía, a Crossing

Nation-states are using geography strategically to deny people access to asylum policies and other legal and human rights. Through voids in policymaking and loopholes in laws, states produce and then manage crises, enabling the strategic use of geography to restrict access to asylum. They are especially—though by no means exclusively—implementing exclusionary practices in extraterritorial locales. These practices become possible through corresponding geographies and discourses that criminalize and exclude asylum seekers; they position, mark, and homogenize the bodies of transnational migrants in particular ways.

It would be a different endeavor to write a book like this one about Canada or about Chinese migration, to write solely from sites of particularity and spaces of exception. Too much scholarship on migration does precisely this. Yet neither Canadian enforcement nor Chinese migration by sea proves exceptional, as international examples have shown. Agamben helpfully names the preponderance of such moments of crisis as exceptionalism. His point is to write against spaces of exception, against the violence perpetrated by states in these times and places too easily labeled "exceptional." I have aimed here to connect global trends and practices with particular incarnations and expressions rooted in local histories and political struggles.

State enactments of international borders sit on a precipice where the new citizen is included through exclusion. Globally, national borders are drawn hastily and haphazardly around the body: they are biopolitical borders. Without a doubt, states have turned their attention to the containment of the body: incarceration rates rise exponentially within the United States; the U.S. model of imprisonment is exported abroad; and Canadians and others debate where, when, and whom to detain. The body becomes a site of global contestation, of struggles to resist the effects of the state and to be heard and represented.

When states and human smugglers construct tunnels, individuals are resubjectified in the spaces between states. The body is contained and excluded, at once inside and outside of the state. In response, detainees resist with hunger strikes and suicide attempts, highlighting the direct relationship between state strategies of containment and bodily practices of resistance.

The thresholds where migrants find themselves are intentionally ambiguous. States perform excision and exclusion—sometimes quietly and other times in forceful, performative fashion for large, politicized audiences. Some stories will be offered up to the media; others will never be told. Human smuggling events divert attention from irresolvable, protracted refugee situations where "durable solutions" to long-term displacement remain elusive. Genuine protection needs and refugee crises are overshadowed by fears of migrant invasions in the form of boat arrivals. Crises in the detention centers on Guam, the Canary Islands, Lampedusa, Melilla, and Nauru draw the state out of its safe bureaucratic offices and into the limelight of crisis. Suddenly civil servants find themselves on the front pages of newspapers, on national nightly news, on live news feeds of interceptions at sea, and at the center of public debate. Crises along borders—at sea, land, and air crossings—intensified throughout the 1990s and gave way to the antiterrorist security walls constructed after 9/11. In this securitized climate, the asylum seeker who arrived spontaneously featured as a criminalized security threat.

The underresourced and unprepared state that refuses to design sustainable policy solutions enters willingly into the performance of crisis at its borders. This is the defensive state, responding and reacting to boats and other threats approaching its borders. Yet this is simultaneously the state that performs on the offense. Policy on the fly gives way to detention, processing in remote locales, and the contracting out of asylum to suprastate institutions, poorer states, and private companies. Boat arrivals construed as crises move securitized policy agendas forward. As Louise Amoore (2007) argues, vision and visuality become key registers through which sovereignty is enacted.

Following crises at coastal and land borders, states extend their borders and turn themselves out on the offense with tactical operations abroad. They network to share information and resources, to collaborate with intelligence and chartered flights of removals. Crisis precipitates action; reaction precipitates preemptive action; human smuggling precipitates border enforcement; terrorist act precipitates antiterrorism security framework. Caught up in these processes through countless modes of exclusion is the person seeking asylum.

Border enforcement projects perpetuate the myth that global migration can be tamed, controlled, "managed" in the name of national security. Some of us live out the myth of securely belonging to multicultural societies, while others witness the violence of exclusion.

How will these alternative states of migration be reconciled? Some of us inhabit the world of policy-speak: the "spontaneous arrival," "voluntary return," "temporary protection," "reintegration package." Others inhabit the fear of alarmist rhetoric: the bogus refugee, the refugee crisis, the disease, the security threat, the "asylum shopper" who exercises choice. Still others fight for a multicultural future where multiple nationalities are possible in societies that promise "integration" and upward mobility.

In this quote from her classic text *The Borderlands / La Frontera*, Gloria Anzaldúa connects knowledge to location, asking us to contemplate *who* we are and *where* we are located in relation to borders.

> Every increment of consciousness, every step forward is a *travesía*, a crossing. I am again an alien in new territory. And again and again. But if I escape conscious awareness, escape "knowing," I won't be moving. Knowledge makes

me more aware, it makes me more conscious. Knowing is painful because after "it" happens I can't stay in the same place and be comfortable. I am no longer the person I was before. (1987, 48)

What do we do with the knowledge set forth in the book about migration? What new stories can be told?

Geographies of smuggling and interception continue to expand where additional states of migration mimic practices outlined in this book. Thai and Israeli officials have made international news with marine interceptions. Meanwhile, political change occurs in many of the countries discussed, introducing eras of flux and uncertainty in states of migration for those seeking asylum. Anti-immigrant, reactionary governments hold political power in a number of states in the European Union. Australian Prime Minister Kevin Rudd announced an end to the Pacific Solution during his first months in office. It eventually emerged, however, that its residual effects remained and intensified offshore, with the detention facility on Christmas Island filled beyond capacity and the construction of new facilities on Indonesian islands. In the United States, President Barack Obama promised change yet studiously avoided the politically charged topic of immigration reform during his first year in office, as residual enforcement infrastructure remained: continued construction of new detention facilities, controversial deportations, and the intense criminalization of immigrant populations in the form of raids, arrests, and local enforcement.

In October 2009, Canadian authorities intercepted the first boat to arrive since the arrivals from Fujian ten years before. This ship carried seventy-six Sri Lankan Tamil men. After being intercepted, the men were placed in detention in Surrey, a suburb of Vancouver and denied release, their arrival associated with security concerns about the LTTE in Sri Lanka (*Globe & Mail* 2009). Simultaneously, similar vessels carrying Sri Lankans were intercepted by the Australian Navy and towed to Indonesia. Passengers on one ship started a hunger protest and refused to disembark. They appealed to the global community for help and communicated via mobile phone with relatives in Canada, rather than concede to the detention that they knew awaited them. "We want a resolution from any foreign country that says they are willing to take us. . . . Nobody wants to come off the boat," stated one asylum seeker named Alex (Gartrell 2009). Their presence on the ship off western Java represented most poignantly the shrinking spaces of asylum, with nowhere to go but into detention.

The challenges confronting states of migration will not be solved simply, partially, or in times of crisis. Nor will they be solved by shifting the borders of sovereignty. We currently inhabit a frightening threshold, between a time, only six decades ago, when asylum was offered as a means to protect those displaced by war to a time when the selective movement of particular migrants is facilitated while far too many others are abandoned. States continue to shrink spaces of protection through creative geographical tactics. Viewed collectively, these are *not* exceptional circumstances prompting unusual responses, but rather the norm. How many long tunnels can be routed and rerouted through these mazes of (in)security that bring borders so violently to the threshold of the body? What do we do with this knowledge, and what state will we be in?

States continually attempt to bring order to the disorder that is human migration. The popular language of "migration management" expresses this effort. Policymakers and politicians will invariably opt for managed over spontaneous migration. But displacement defies management and is, by its very nature, a spontaneous act of dispossession for those displaced.

There are basic principles of asylum that states sometimes overlook in their response to human smuggling: that it is not illegal to make an asylum claim, that such claims should be allowed rather than prohibited, that histories must be heard and assessed individually, and that such incidents need not and should not be construed as crises. People traveling on boats, held in detention, and being processed in airports have distinct identities, histories, and desires that need to be heard. They have a right to seek asylum.

Notes

Introduction

1. Those migrants who arrive spontaneously in Canada and make refugee claims are called "refugee claimants." Those who seek protection once reaching sovereign territory in the United States or Australia are "asylum seekers."

2. After the terrorist attacks in the United States in 2001, Citizenship and Immigration Canada remained intact, but the mandate to enforce borders was moved to a new agency called the Canadian Border Services Agency.

3. This field research saw many twists and turns. See Mountz (2007) for a fuller discussion.

4. Within CIC and other organizations, a small group of people were most involved in the response to these arrivals, which makes them easily identifiable. Pseudonyms are used and identifying details sometimes altered to conceal identities of participants. I specify the general location but not the specific dates of interviews due to the public nature of the schedule of some officials interviewed.

5. Often analyses of the contemporary political landscape operating under the rubric of globalization emphasize movement, flows (Appadurai 1996; Castells 2000), "the information society" (Castells 1996), identities, scales (Brenner 1997), ethnoscapes (Appadurai 1996), social movements (Guarnizo and Smith 1999), and networks (Agnew 1999) that supersede an ever-present backdrop of the more statically imagined nation-state. Still others see nation-states not as excluded by these networks, but rather implicated and embedded within them (Castells 2000; Hardt and Negri 2000; Taylor 2000). Saskia Sassen, for example, argues that nation-states are not dead but rearticulating, and she spatializes these rearticulations in the particular locales of body and border (1996, 2006). Arjun Appadurai (2006, 129), alternatively, offers the metaphor of the vertical vertebrate.

6. Understanding the ways in which nation-states see, classify, accept, or reject transnational migrant subjectivities shows the power of the state to produce identity and to thus materially reject those not included, as in Nevins's analysis of the construction of the "illegal alien" (2002) in the United States, which Heyman asserts

is a key context to the organizational worldviews of INS officers (1995). The structural effect has an important impact on our understanding of border enforcement. Joe Nevins (2002) studied the history of U.S. enforcement of its border with Mexico. Nevins (2002, 160) argues that those practices that reify the artificial boundary between state and civil society are depoliticizing because they assume that nation-states act as autonomous decisionmakers. When people believe that states are all-powerful and mysterious, existing somewhere "out there," they do not participate in protest or dialogue.

7. Kristof Van Impe (2000, 120) suggests that trafficking is the "degeneration" of smuggling, signaling a loss of control on the part of clients who cannot pay off debts. The distinction between these terms remains ambiguous—more a difference of degree than a dichotomy—and is complicated by the fact that so little is known about the conditions under which individuals are smuggled or trafficked and the conditions in which they ultimately live and work at the final destination (see Kwong 1997; Chin 1999).

8. Contemporary research on transnational migration often foregrounds migrant narratives to theorize dynamic subjectivities. These are interventions into what were once primarily structural macronarratives of immigration (e.g., Silvey and Lawson 1999; Lawson 2000; McHugh 2000). Because they focus on movement across borders, theorists tended to write about human mobility in celebratory fashion, wherein movement was a triumph over the constraints of borders (see Mitchell 1997a for a critique). They thus tended to overlook the structural impediments to mobility in a postcolonial, postnational era (see Appadurai 1996; Anderson 2000) of quick border crossings and dual citizenship.

1. Human Smuggling and Refugee Protection

1. The practice of turning away boats calls into question the integrity of the 1951 Convention Relating to the Status of Refugees and the 1967 Protocol, to which Canada is a signatory, a point to be explored further in chapter 5.

2. Fears of direct and indirect travel from particular regions have historical precedents in Canada. In 1908 the Canadian government passed the Continuous Journey stipulation to prohibit Indians who stopped in Hong Kong from entering Canada.

3. Trends in scholarship on U.S. borders reflect these gaps in knowledge. Although there is a substantial body of work on the U.S.–Mexico border (e.g., Andreas 2000; Nevins 2002), much less has been written by United States–based scholars about the forty-ninth parallel and Canadian immigration and refugee policies. See Koslowski (2004) and Sparke (2006) for discussions of security along the Canada-U.S. border.

4. These numbers were dynamic and contested. One document released to the public by CIC placed the number of refugee claimants at 549 (CIC, Marine Arrivals: Status Update, 18 February 2000). Their legal status changed over time. Some migrants misrepresented their ages, and minors were identified distinctly by the federal and provincial governments.

5. One million immigrants to Canada passed through Pier 21, Canada's gateway on the east coast, from 1928 to 1971.

6. Among the reasons Canada does not detain refugees more broadly is the successful antidetention lobbying undertaken by its refugee advocacy community, organized under the umbrella group the Canadian Council for Refugees and smaller regional groups located in larger cities such as Vancouver, Montreal, and Toronto.

7. The program was largely dismantled when the Liberal Party came into power in British Columbia in 2001.

8. Following the highly publicized Australian interception of the M.V. Tanu, for example, Prime Minister John Howard refused to allow refugee claimants to be transferred to sovereign territory. Migrants were detained on Christmas Island and Nauru until New Zealand and other countries offered to resettle many of the families (see Mares 2002; Perera 2002b).

2. Seeing Borders Like a State

1. The use of vision and sight as metaphors for how states see needs to be further problematized, as they are not inclusive terms and certainly do not represent how many people experience landscapes.

2. These were small coastal towns with docks in Alaska and British Columbia, including Ucluelet, Port Renfrew, Port Alberni, Tofino, Gold River, Tahsis, Zeballos, and Port Alice. At each location CIC employees identified the infrastructure that would be necessary to carry out operations, such as housing for civil servants, sites for temporary processing and detention, roads to transport migrants inland, medical clinics, helicopter pads, and runways.

3. In the Canadian Supreme Court, the Korean crew argued that the ship had been pirated by smugglers and they were forced to deliver the migrants. They were eventually acquitted.

4. Whereas "snakehead" refers to the human smuggler leading the operation, "enforcers" are those employed to accompany migrants on the journey. They are often people with histories of gang involvement who take the job in exchange for the fare for the trip (Chin 1999).

5. Tests screening for hepatitis B, parasites, syphilis, and other sexually transmitted diseases were conducted at the discretion of physicians.

6. Consultants subcontracted by CIC suggested that some images were linked to triads.

7. Ko Lin Chin (1999) is among those who believe that the involvement of transnational organized crime is a myth. He suggests, rather, that many Chinese smuggling movements are "one-offs": the entrepreneurial confluence of interests and opportunities.

8. Early investigative questions addressed the point in the journey where ship captains learned the coordinates of their destination, and the extent to which the

enforcers and clients knew their whereabouts, final destination, and the contacts and procedures that were to ensue upon arrival.

9. Fouad Ajami made this statement as a guest on *Charlie Rose*, 13 September 2001, on the Public Broadcasting Company (PBS).

10. During Prime Minister Campbell's tenure in 1993, enforcement was also separated from facilitative work on immigration to Canada for four months.

11. The financial struggles continued well after the initial response to the arrivals in 1999 and resulted in an internal study in 2002 of "lessons learned" during this response, which bureaucrats saw as an effort to investigate and explain the cost of the response.

12. "Boating season" runs roughly from May to October, after which time weather makes the Pacific too rough.

13. A similar coastal watch program developed in the United States after the attacks on 9/11.

14. CIC and the UNHCR argued in response that legal counsel and refugee advocates who encouraged claimants to exercise every opportunity to appeal negative decisions when the outcome was unlikely to change were themselves to blame for the extended period of detention.

15. Complicating this strategy is the status of minors and families in detention. The UNHCR was particularly interested in this issue. While Australia and the United States detained minors, Canada did not, particularly under the careful watch of the Province of BC, legally the guardian for unaccompanied minors.

16. This strategy repeated a campaign undertaken by the United States in 1993 (P. Smith 1997, 4) and 2000 (*New York Times* 2000b), and in Australia in 2001.

17. They were ultimately acquitted. Other enforcers, however, were convicted of smuggling and sentenced to four years in prison (*Globe & Mail* 2001a).

3. Ethnography of the State

1. The call centers of CIC are a classic geographical representation of this reality. Applicants are not able to call the office where their application is being processed, but rather must contact call centers that are geographically detached from sites of application or processing. At the call center, a representative accesses application status information via a centralized computer system.

2. Three studies by geographers examine the governance of mobility and enforcement, displacement, and territoriality (Herbert 1997; Hyndman 2000; Nevins 2002) at different scales. Jennifer Hyndman (2000) conducted an institutional analysis of the United Nations High Commissioner for Refugees (UNHCR). With refugee camps in Kenya and Somalia as field sites, her work uncovers geographical relationships between refugees and states, and the intermediary management role of the UNHCR (2000). Steve Herbert (1997) conducted fieldwork with the Los Angeles Police Department and did ride-alongs to study the construction of the neighborhood. He tests the relationship between human agency and structural constraints such as the law (1997).

3. Researching the daily operation of a science laboratory in England, John Law theorized the lab not as one site but as a "pastiche" of different places (1994, 40). Nyers's (2006) notion of incompletion echoes Law's conceptualization of the organization as a network that is never pure and never truly ordered, but rather is a perpetual "dream" of ordering: "ordering that is never complete, and runs at cross purposes in a hundred different occasions, and never adds up to the hideous purity of an order—even though it generates a set of processes that we can call 'The Lab'" (1994, 39).

4. All managers had two phone lines: one listed on a business card but rarely answered and a cell phone, which they did tend to answer. This strategic use of telephone lines seemed to signal the sincerity of their intention to communicate.

5. After a struggle among employees about whether or not I should be added to listservs, the IT staff hooked me up to a computer with its own dial-in line to the Internet.

6. The department experienced another major internal reorganization and reduction in staff with the placement of its enforcement mandate and staff in the Canadian Border Services Agency in December 2003.

7. Estimated expenditures on detention and removals alone in the response to the boat arrivals far exceeded the annual operating budget for the entire region at approximately $36 million.

8. To contextualize this investment in human resources, there were only three people working in intelligence in the entire BC–Yukon region, only one of whom was asked to gather intelligence about smuggling movements.

9. This is close to when Ottawa's workday starts, taking into account the three-hour time difference.

10. The Canadian Broadcasting Corporation (CBC) is the public-channel equivalent to the Public Broadcasting System (PBS) in the United States and the British Broadcasting Corporation (BBC) in the United Kingdom. CTV Newsnet is the corporate Canadian equivalent to CNN in the United States.

11. This is an interesting twist. States are often thought to be the agents, rather than the objects, of surveillance; but these findings correspond with the overall feeling that bureaucrats expressed in interviews, that they often feel quite the opposite—in the dark.

12. This was especially true of processing at the base in Esquimalt. When I began research in 2000, managers were working on procedures for the expected arrivals for the next summer. "No, it was just all up here, trying to remember everything that we did last year. You know, for procedures, just trying to think. So mostly, it was just all in my head, what didn't work last year."

13. This statement was made as claimants made their way through the refugee determination process.

14. Michael Herzfeld calls time "the great social weapon" of the bureaucracy (1992). At CIC, everyone worked within a specific time frame, among different institutions and even among those people working within different branches. In the

response to boat arrivals, the time that it takes to conduct an operation differs from the time to gather intelligence, to conduct an investigation, for communications to respond to the media, for human resources to shuffle people around, and for lawyers respond to a legal issue.

4. Crisis and the Making of the Bogus Refugee

1. Ghassan Hage refers to phrases such as "too many" as racialized "categories of spatial management" (1998, 38).

2. Mahtani and Mountz (2002) found that newspapers paid significant attention to criminal associations with immigration. The newspapers frequently contributed to a collective criminalization of refugee applicants in particular. This criminalization occurred through the linking of racialized groups to categories of crime. Two recurring examples in BC include the association of Asian immigrants with gang-related crime and of Latino immigrants with drug-related crime. The association of criminality with racialized immigrant groups in Vancouver increased in 1998, setting the stage for criticism of refugee policies with the boat arrivals in 1999.

3. It is interesting to think about this language as an Americanization and, therefore, harmonization of borders through discourse.

4. One lawyer interviewed expressed frustration with the media and noted that representations made their way into the detention reviews, where federal lawyers showed pictures of rusty boats as part of the process of identifying this group. This lawyer asked an important question: "What does the rust have to do with it?" (Interview, Vancouver, September 2001)

5. These images reflect similar constructions of Chinese migrants as harbingers of disease during migrations to Canada a century earlier (Anderson 1991; Li 1998).

6. More accurately, the province of British Columbia entered the business of detaining refugee claimants by way of a lucrative contract with the federal government.

7. See Rajaram (2002) for a discussion of the notion of the refugee as helpless.

8. There is a frequent and contentious legal debate in the area of interception at sea: whether, when, how, and in what language do authorities allow conditions for an asylum claim.

9. This term refers to the Greater Vancouver area.

10. The Legal Services Society is a provincial body funded by the provincial government.

11. The Refugee Committee of the BC branch of the Canadian Bar Association wrote a letter to the Legal Services Society concerning the bidding process and the award of a contract to one lawyer in particular.

12. Lawyers detailed the ways in which access to clients in detention was limited. Because of the challenge of finding skilled interpreters with security clearance from the RCMP and because of guard breaks during the day, they spent a lot of time

waiting to meet with clients and, according to them, less time than normal meeting with clients to prepare their Personal Information Form. This is the primary document that must be filed by the refugee claimant with the IRB within twenty-eight days of making a claim. It solicits information on biographical details relevant to the refugee claim. "The whole preparation was done in an accelerated fashion as well; because these people were detained, the Immigration and Refugee Board gave priority to having their claims heard. So the regular process that you would have to sit down, meet with a client for three, four, five times with a certified interpreter to prepare their statements was out. Lawyers were preparing the forms in less time and without the qualified interpreters, and with claimants who didn't really know if they could trust you or not" (Charlton et al. 2002, 15). The difficulty of assessing some of these arguments relates to the number of institutions involved. The IRB faced significant logistical challenges in processing this group. BC Corrections had limited experience working with refugee claimants, and CIC faced criticism for the lengthy detention.

13. The term "Gold Mountain" refers to the movement of Chinese immigrants to North America in the mid-1800s (originally to California and soon thereafter to Canada) in search of gold.

14. This argument echoes frequent insinuations by refugee counsel that the "arm's-length" relationship between CIC and the IRB is not adequately maintained. The Minister of Immigration appoints members of the IRB and determines renewal of their contracts. (Interview, Vancouver, September 2001)

15. One lawyer recounted the experience of representing a client on a Friday afternoon before members of the IRB who had already been working in Prince George on a temporary basis for a week or two. As it began to snow outside, he said, they rushed out of the hearing to take a flight home to Vancouver. (Interview, Vancouver, August 2001)

16. In 1999 the average refugee determination process lasted nine months (United States Committee for Refugees 2000a), so not all of the hearings moved faster than the average for such hearings, although some certainly did.

17. There is some numerical discrepancy here. The United States Committee for Refugees (2000a) cites a 2.5 percent acceptance rate among the Fujianese migrants who arrived by boat and made refugee claims, whereas the ratio of 24 out of some 549 claims comes out to around 4.5 percent. Perhaps this discrepancy relates to the challenges of quantifying the legal status of a dynamic population, causing different sources to cite different numbers of total refugee claimants among this group. An article published in the *Vancouver Sun* (2009) confirms 24 acceptances out of a total of 577 claims (the former number has remained consistent; the latter has changed over the years). This is equivalent to a 4.2 percent acceptance rate.

18. Over time, some 330 migrants were repatriated (United States Committee for Refugees 2001). According to the estimates of a Vancouver-based refugee lawyer who represented claimants, some 300 migrants were represented by lawyers hired

by Legal Aid through the bidding process (see Mountz 2003). Approximately 100 were not covered through these contracts. Twenty-two of the positive claims were among the 100 not represented by contracts awarded through bids, and most of them were located in Vancouver. (Interview, Vancouver, September 2001.)

19. Ellis and Wright (1998), Cresswell (1997, 2006), and Sibley (1995) have analyzed spatial metaphors of displacement that deem bodies out of place. Some bodies are more visible because of race, class, gender, and citizenship, all of which figure prominently in discourse on immigration and are central to decisions about who "belongs" to the nation-state.

20. This is a reference to Allen Pred's (1995) book *Even in Sweden*, where he argues that even in Sweden, racism toward immigrants takes place.

21. This is an interesting parallel to Nelson's analysis of the Guatemalan state as a wounded body and of ethnic organizing as "the finger in the wound" (1999). She reviews countless images of the ways in which the nation-state in turn positions the bodies of Guatemalan citizens through articulations of national identity.

22. Singular narratives of the state are masculinist projects; the maintenance of the solid, secure, uniformed, and unified bodies of officials is a way to preserve the power of the message. Within political geography, this points to a need to challenge the securities of what Jessop (1990) calls "strong" state theories.

23. Feminist scholars have intervened in debates about globalization and migration to argue that certain scales have been overlooked (Marchand and Runyan 2000; Marston 2000). Nagar and colleagues (2002) suggest that a focus on power at various scales is central to feminist interventions into the "subjects and spaces of globalization." Shifting to the scale of the body, they note that "starting from the standpoint of people and economic spheres that are marginalized under capitalist processes reveals the ways in which contemporary globalization is intimately tied to gendered and racialized systems of oppression" (2002). As Sassen suggests, while the nation-state loses power in some realms, it reasserts power at sites such as the border and the body (1996, 65). Reading the state through the body counters the argument that the state has met its demise in a borderless (Ohmae 1995), "post-national" world (Soysal 2000). These calls for new scalar narratives are alive in a feminist geopolitical framework that "relies upon a commitment to the safety and well-being of persons rather than states" (Hyndman 2001, 219).

5. Stateless by Geographical Design

1. I see these graduated degrees of statelessness as one way of spatializing Aihwa Ong's notion of "graduated zones of sovereignty" (1999), wherein populations have very different relationships with states and corresponding degrees of mobility.

2. In the United States and Australia, migrants who arrive "spontaneously" and make a claim for protection from abroad are categorized "asylum seekers." In Canada, these are "refugee claimants." All three countries select and resettle refugees from abroad.

3. Typologies are always provisional, a way to organize a polyglot of practices.

4. Of course this is an important distinction, particularly for legal reasons. Increasingly, however, nation-states are treating "mixed flows" as homogeneous groups that are not Convention refugees.

5. This waffling related, in Canada, to an ongoing debate about the ownership of the mandate to respond to human smuggling. In fact, these struggles caused the delay of policy for a few years following the 1999 interceptions, during which time federal departments drafted and then debated an agreement regarding interdepartmental plans to intercept marine arrivals. Whereas CIC held primary responsibility in 1999, this responsibility was handed over when the enforcement mandates of immigration were separated from the facilitative mandate with the establishment of the CBSA.

6. This changed in 2008, when the passage of an omnibus bill pulled the other islands back under the jurisdiction of U.S. immigration law.

7. States and corporations are also reconfiguring borders with technology. In early June 2004, Accenture, a private company based outside the United States, was awarded a $10 billion dollar contract by the U.S. government to create "virtual borders." The program did cement and extend the biometric technologies developed to regulate transnational mobility by recording the fingerprints of all visa applicants in a global digital database. This database tracks biometrically enabled smart cards with the intention of monitoring over 400 land, air, and sea ports of entry (Accenture 2004). We need, therefore, to look not only at the intersections between geography and the law, but also at the role of technology in the privatization of enforcement (Amoore 2006).

6. In the Shadows of the State

1. Temporary Protected Status provides temporary protection to those applying for asylum in the United States while their cases are adjudicated. This program has recently been available to people from Burundi, El Salvador, Honduras, Liberia, Montserrat, Nicaragua, Somalia, Haiti, and Sudan (see Wasem and Ester 2004).

2. Despite the end of the cold war, the United States still favors asylum applicants from communist states.

3. Although he began by analyzing the histories of institutions, such as penal systems, hospitals, and asylums, Foucault's project is more generalizable to the conceptualization of power (1995, 216). His ideas provoke analysis not only of institutional structures and laws, but also of the power of the state *beyond* its apparatus (1980). Hannah (1997, 344) emphasizes the need to push Foucauldian analyses beyond institutional frameworks but is critical of applications beyond institutions that tend toward "motif," "mood setter," or "vague guide" (1997, 344).

4. Pseudonyms are used to protect the identities of participants.

5. This work responds to various critiques of Foucault's work and its applications, including the failure to address the state (e.g., Driver 1985, 438–39), to conduct analyses beyond the spaces of institutions (Hannah 1997), and to allow for human agency.

6. Mitchell suggests retheorizing connections between the local and the global (1997b, 110), and between more abstract discussions of borders and more "grounded" material realities (106). Similarly, Guarnizo and Smith (1998) advocate research at multiple scales with simultaneous macro- and microanalyses for a "grounding" of transnationalism that does not romanticize migrant narratives (24). It is theoretically insipid, however, to conceptualize the state, transnational corporations, and supra-state bodies as structure and the migrant as agency. Mahler (1998) has critiqued transnationalism from "above" and "below" as a binary that does not account for complex subjectivity through which one individual may in fact be an oppressed worker in one national context and an oppressive employer in another (65).

7. At the federal immigration building in downtown Newark, officers closed the street to traffic to improve security, rendering those who wait in line outside for hours visible.

8. The ambiguities of policy require most immigrants interacting with the federal government to hire lawyers. An exploding economy of artificial legal services thrives wherever there exist potential clientele (Mahler 1995). Communities of immigrant service providers also advocate for policy changes and interpret and negotiate policy for immigrants (Coutin 2000).

9. These include Temporary Protected Status, Deferred Enforced Departure, the American Baptist Church Agreement, and the Nicaraguan and Central American Relief Act. Expiration dates for these programs were extended multiple times through lobbying and legal procedures. See Coutin (1998; 2000), Menjívar (2000), and Mahler (1995) for discussion of the history and implementation of these policies.

10. Geographers Adrian Bailey, Ines Miyares, Richard Wright, and I conducted research and participated in the political struggle for residency between 1997 and 1998. We conducted eighty-five interviews with Salvadoran asylum applicants and their families in northern New Jersey and in El Salvador, and with the service providers and advocates assisting them.

11. See Heyman 2007 for a discussion of the relationship between immigrant rights movements and lived experiences of migration.

12. This case draws on extensive interviews and participant observation with undocumented Mexican residents in Poughkeepsie, New York, and their families in Oaxaca between 1993 and 1995 (Mountz 1995).

13. The most poignant example occurred one weekend morning when there was a funeral march in San Agustín. I stopped by the house of a family I had planned to visit. They had been peering out their door when the march started, wondering who had died in San Agustín the night before, when their brother called from Poughkeepsie to explain what had happened.

14. In 1999, volunteers from W. W. Smith Elementary School in Poughkeepsie installed the first computer in the school in San Agustín.

Bibliography

Abrams, P. 1988. "Notes on the Difficulty of Studying the State" (1977). *Journal of Historical Sociology* 1, no. 1: 58–89.

Accenture. 2004. "U.S. Department of Homeland Security Awards Accenture-Led Smart Border Alliance the Contract to Develop and Implement US-VISIT Program," news release, 1 June 2004. www.accenture.com/xd/xd.asp?it=enweb&xd=_dyn%5Cdynamicpressrelease_730 (accessed 24 June 2004).

Agamben, G. 1998. *Homo Sacer: Sovereign Power and Bare Life*. Stanford: Stanford University Press.

———. 2005. *States of Exception*. Chicago: University of Chicago Press.

Agnew, J. 1994. "The Territorial Trap: The Geographical Assumptions of International Relations Theory." *Review of International Political Economy* 1: 53–80.

———. 1999. "Mapping Political Power beyond State Boundaries: Territory, Identity, and Movement in World Politics." *Millennium* 28: 499–521.

Amoore, L. 2006. "Biometric Borders: Governing Mobilities in the War on Terror." *Political Geography* 25, no. 3: 336–51.

———. 2007. "Vigilant Visualities: The Watchful Politics of the War on Terror." *Security Dialogue* 38, no. 2: 215–32.

Amoore, L., and de Goede, M., eds. 2008. *Risk and the War on Terror*. New York: Routledge.

Anderson, K. 1991. *Vancouver's Chinatown: Racial Discourse in Canada, 1875–1980*. Montreal, Kingston: McGill-Queen's University Press.

———. 2000. "Thinking 'Post-Nationally': Dialogue across Multicultural, Indigenous and Settler Spaces." *Annals of the Association of American Geographers* 90, no. 2: 381–91.

Andreas, P. 2000. *Border Games: Policing the U.S.-Mexico Divide*. Ithaca, N.Y.: Cornell University Press.

———. 2003. "A Tale of Two Borders: The U.S.-Mexico and U.S.-Canada Lines After 9/11" (working paper, Center for Comparative Immigration Studies). http://repositories.cdlib.org/ccis/papers/wrkg77 (accessed 10 October 2003).

Anzaldúa, G. 1987. *Borderlands/La Frontera: The New Mestiza.* San Francisco: Spinster/Aunt Lute Press.

Appadurai, A. 1996. *Modernity at Large: Cultural Dimensions of Globalization.* Minneapolis: University of Minnesota Press.

———. 2006. *Fear of Small Numbers: An Essay on the Geography of Anger.* Durham, N.C.: Duke University Press.

Augé, M. 1995. *Non-Places: Introduction to an Anthropology of Supermodernity.* New York: Verso.

Bailey, A., R. Wright, A. Mountz, and I. Miyares. 2002. "Producing Salvadoran Transnational Geographies." *Annals of the Association of American Geographers* 92, no. 1: 125–44.

Bakhtin, M. 1981. *The Dialogic Imagination: Four Essays.* Austin: University of Texas Press.

BarTalk. 1999. "The Right—and Cost—of Representation: New Tide of Migrants Raises Legal and Policy Issues. 11, no. 5: 1.

Basch, L., N. Glick Schiller, and C. Szanton Blanc. 1994. *Nations Unbound: Transnational Projects, Postcolonial Predicaments and Deterritorialized Nation-States.* Amsterdam: Gordon and Breach.

Bashford, A., and C. Strange. 2002. "Asylum-Seekers and National Histories of Detention." *Australian Journal of Politics and History* 48, no. 4: 509–27.

BBC News. 2009. "Migrants Escape on Italian Island." http://newsvote.bbc.co.uk/ mpapps/pagetools/print/news.bbc.co.uk/2/hi/europe/7848786.stm (accessed 25 January 2009).

———. 2010. "Analysis: Europe's Asylum Trends." http://news.bbc.co.uk/2/hi/ europe/4308839.stm (accessed 4 January 2010).

Beiser, M. 1999. *Strangers at the Gate: The "Boat People's" First Ten Years in Canada.* Toronto: University of Toronto Press.

Betts, A. 2004. "The International Relations of the "New" Extraterritorial Approaches to Refugee Protection: Explaining the Policy Initiatives of the UK Government and UNHCR." *Refuge* 22, no. 1: 58–70.

Bhabha, J. 2005. "Trafficking, Smuggling, and Human Rights." *Migration Information Source* March: 1–5. http://www.migrationinformation.org/Feature/display .cfm?id=294 (accessed 2 January 2010).

Bhandar, D. 2008. "Resistance, Detainment, Asylum: The Ontological Limits of Border Crossing in North America." In *War, Citizenship, Territory,* edited by D. Cowen and E. Gilbert, 281–302. New York: Routledge.

Bigo, D. 2002. "Security and Immigration: Toward a Critique of the Governmentality of Unease." *Alternatives* 27: 63–92.

Bloch, A., and L. Schuster. 2005. "At the Extremes of Exclusion: Deportation, Detention and Dispersal." *Ethnic and Racial Studies* 28, no. 3: 491–512.

Blomley, N. 1994. *Law, Space and the Geographies of Power.* New York: Guilford Press.

Boswell, C. 2003. "The External Dimension of EU Immigration and Asylum Policy." *International Affairs* 79, no. 3: 619–38.

Bowden, C. 2003. "Outback Nightmares & Refugee Dreams." *Mother Jones* March/April. http://motherjones.com/politics/2003/03/outback-nightmares-refugee-dreams (accessed 2 January 2010).

Brenner, N. 1997. "State Territorial Restructuring and the Production of Spatial Scale." *Political Geography* 16, no. 4: 273–306.

Brenner, N., B. Jessop, M. Jones, and G. MacLeod, eds. 2003. *State/Space*. Malden, Mass.: Blackwell.

Brouwer, A. 2003. "Attack of the Migration Integrity Specialists! Interdiction and the Threat to Asylum." http://www.web.net/%7Eccr/interdictionab.htm (accessed 15 November 2003).

Burawoy, M., ed. 2000. *Global Ethnography: Forces, Connections, and Imaginations in a Postmodern World*. Berkeley: University of California Press.

Butler, J. 1990. "Performative Acts and Gender Constitution: An Essay in Phenomenology and Feminist Theory." In *Performing Feminisms: Feminist Critical Theory and Theatre*, edited by S. E. Case, 270–82. Baltimore: Johns Hopkins University Press.

———. 1993. *Bodies That Matter: On the Discursive Limits of Sex*. New York: Routledge.

———. 2004. *Precarious Life: The Powers of Mourning and Violence*. New York: Verso.

Butler, J., and G. Spivak. 2007. *Who Sings the Nation-State? Language, Politics, Belonging*. Calcutta: Seagull Books.

Calavita, K. 1992. *Inside the State: The Bracero Program, Immigration, and the I.N.S.* New York: Routledge.

Campbell, D. 1992. *Writing Security: United States Foreign Policy and the Politics of Identity*. Manchester: Manchester University Press.

Canadian Council for Refugees. 2003. "Interdiction and Refugee Protection: Bridging the Gap." Proceedings, International Workshop, 29 May 2003, Ottawa.

Carter, D. 1997. *States of Grace: Senegalese in Italy and the New European Immigration*. Minneapolis: University of Minnesota Press.

Castells, M. 1996. *The Rise of the Network Society. The Information Age: Economy, Society, Culture*, vol. I. Malden, Mass.: Blackwell.

———. 2000. *End of Millennium. The Information Age: Economy, Society and Culture*, vol. III. Malden, Mass.: Blackwell.

Castles, S., and N. Van Hear. 2005. *Developing DFID's Policy Approach to Refugees and Internally Displaced Persons*. Oxford: Refugee Studies Centre, University of Oxford.

Chang, K., and L. H. M. Ling. 2000. "Globalization and Its Intimate Other: Filipina Domestic Workers in Hong Kong." In *Gender and Global Restructuring: Sightings, Sites and Resistances*, edited by M. Marchand and A. Runyan, 27–43. New York: Routledge.

Charlton, A., S. Duff, D. Grant, A. Mountz, R. Pike, J. Sohn, and C. Taylor. 2002. "The Challenges to Responding to Human Smuggling in Canada: Practitioners Reflect on the 1999 Boat Arrivals in British Columbia." Research on Immigration and Integration in the Metropolis Working Paper No. 02-23. http://www.riim.metropolis.net (accessed 20 January 2003).

Chin, K. L. 1999. *Smuggled Chinese: Clandestine Immigration to the United States.* Philadelphia: Temple University Press.

Cho, G. 2008. *Haunting and the Korean Diaspora: Shame, Secrecy and the Forgotten War.* Minneapolis: University of Minnesota Press.

Citizenship and Immigration Canada. 2000. "Facts and Figures: Immigration Overview." Minister of Public Works and Government Services Canada.

———. 2001. "Review of the Immigration Control Officer Network: Final Report." http://www.cic.gc.ca/EnGLish/resources/audit/ico/index-e.asp (accessed 2 January 2010).

Clark, G., and M. Dear. 1984. *State Apparatus: Structures and Language of Legitimacy.* Boston: Allen & Unwin.

Clarkson, S. 2000. "600 Is Too Many." *Ryerson Review of Journalism* Spring. http://www.rrj.ca/issue/2000/spring/306/ (accessed 2 January 2010).

Clifford, J., and G. E. Marcus. 1986. *Writing Culture: The Poetics and Politics of Ethnography.* Berkeley: University of California Press.

Coleman, M. 2007. "Immigration Geopolitics beyond the US-Mexico Border." *Geopolitics* 39, no. 1: 54–76.

———. 2009. "What Counts as the Politics and Practice of Security, and Where? Devolution and Immigrant Insecurity after 9/11." *Annals of the Association of American Geographers* 99, no. 5: 904–13.

Conquergood, D. 2002. "Performance Studies: Interventions and Radical Research." *Drama Review* 46, no. 2: 145–56.

Coutin, S. 1998. "From Refugees to Immigrants: The Legalization Strategies of Salvadoran Immigrants and Activists. *International Migration Review* 32, no. 4: 901–25.

———. 2000. *Legalizing Moves: Salvadoran Immigrants' Struggle for US Residency,* Ann Arbor: University of Michigan Press.

Cresswell, T. 1997. "Weeds, Plagues, and Bodily Secretions: A Geographical Interpretation of Metaphors of Displacement." *Annals of the Association of American Geographers* 87, no. 2: 330–45.

———. 2006. *On the Move: Mobility in the Modern Western World.* New York: Routledge.

Crisp, J. 1997. "The Asylum Dilemma." In *State of the World's Refugees: A Humanitarian Agenda,* edited by J. Crisp, M. Ronday-Cao, and R. Reilly. Oxford: Oxford University Press.

Cunningham, H., and J. Heyman. 2004. "Introduction: Mobilities and Enclosures at Borders." *Identities: Global Studies in Culture and Power* 11: 289–302.

Dean, M. 1999. *Governmentality: Power and Rule in Modern Society*. Los Angeles: Sage.

De Certeau, M. 1984. *The Practice of Everyday Life*. Berkeley: University of California Press.

Del Casino, V. J., A. J. Grimes, S. P. Hanna, and J. P. Jones III. 2000. "Methodological Frameworks for the Geography of Organizations." *Geoforum* 31: 523–38.

DeStefano, A. M. 1997. "Immigrant Smuggling through Central America and the Caribbean." In *Human Smuggling*, edited by P. J. Smith, 134–55. Washington, D.C.: Center for Strategic & International Studies.

Direct Action Against Refugee Exploitation. 2001. "Movements across Borders: Chinese Women Migrants in Canada." Available from authors at dare_vancouver@hotmail.com and online at http://www.harbour.sfu.ca/freda/reports/daare.htm.

Driver, F. 1985. "Power, Space and the Body: A Critical Assessment of Foucault's *Discipline and Punish*. *Environment and Planning D: Society and Space* 3: 425–46

Ellis, M., and R. Wright. 1998. "The Balkanization Metaphor in the Analysis of U.S. Immigration." *Annals of the Association of American Geographers* 88, no. 4: 686–98.

European Parliamentary Group. 2006. "Lampedusa and Melilla: Southern Frontier of Fortress Europe." European United Left/Nordic Green Left.

Flynn, M., and C. Cannon. 2009. "The Privatization of Immigration Detention: Towards a Global View" (Global Detention Project working paper). http://webprod.iheid.ch/webdav/site/globalmigration/shared/_communIntersites/GDP_PrivatizationPaper_Final5.pdf (accessed 2 January 2010).

Foster, L. 1998. *Turnstile Immigration: Multiculturalism, Social Order & Social Justice in Canada*. Toronto: Thompson Educational Publishing.

Foucault, M. 1980. *Power/Knowledge: Selected Interviews & Other Writings, 1972–1977*. New York: Pantheon Books.

———. 1991. "Governmentality." In *The Foucault Effect: Studies in Governmentality*, edited by G. Burchell, C. Gordon, and P. Miller, 87–104. Chicago: University of Chicago Press.

———. 1995. *Discipline and Punish: The Birth of the Prison*. New York: Vintage Books.

Fuller, G. 2003. "Life in Transit: Between Airport and Camp." *Borderlands E-journal* 2, no. 1: 1–8. http://www.borderlandsejournal.adelaide.edu.au/vol2no1_2003/fuller_transit.html (accessed 24 January 2004).

Gartrell, A. 2009. "Asylum Seekers en Route to Australia Declare Hunger Strike." 16 October 2009. http://www.news.com.au/story/0,27574,26216451-401,00.thml (accessed 18 October 2009).

Gatens, M. 1991. "Corporeal Representation in/and the Body Politic. In *Cartographies: Poststructuralism and the Mapping of Bodies and Spaces*, edited by R. Diprose and R. Ferrell, 79–87. North Sydney: Allen & Unwin.

Gauvreau, C., and G. Williams. 2003. "Detention in Canada: Are We on the Slippery Slope?" *Refuge* 20, no. 3: 68–70.

Gibson-Graham, J. K. 1996. *The End of Capitalism (as We Knew It): A Feminist Critique of Political Economy.* Oxford: Blackwell.

Gilbert, E. 2005. "The Inevitability of Integration? Neoliberal Discourse and the Proposals for a New North American Economic Space after September 11." *Annals of the Association of American Geographers* 95, no. 1: 202–22.

———. 2007. "Leaky Borders and Solid Citizens: Governing Security, Prosperity and Quality of Life in a North American Partnership." *Antipode* 29, no. 1: 77–98.

Glick-Schiller, N., L. Basch, and C. Blanc-Szanton. 1992. *Towards a Transnational Perspective on Migration.* New York: New Academy of Sciences.

Globe & Mail. 2001a. "Chinese Smugglers Get 4 Years." 17 March 2001.

———. 2001b. "Migrants Fare Better by Air. Chinese Landing in B.C. by Boat Treated Worse than Those Arriving by Plane." 14 April 2001.

———. 2009. "Canada to Take Hard Line with Would-Be Migrants." 19 October 2009.

Goffman, E. 1959. *The Presentation of Self in Everyday Life.* Garden City, N.Y.: Doubleday Anchor Books.

Gómez-Peña, G. 1996. *The New World Border: Prophecies, Poems, and Loqueras for the End of the Century.* San Francisco: City Lights.

Goodwin-Gill, G. 1986. "International Law and the Detention of Refugees and Asylum Seekers." *International Migration Review* 20, no. 2: 193–219.

Gordon, A. 2008. *Ghostly Matters: Haunting and the Sociological Imagination.* Minneapolis: University of Minnesota Press.

Gordon, D., and R. Behar. 1995. *Women Writing Culture.* Berkeley: University of California Press.

Gramsci, A. 1994. *Letters from a Prison*, vols. I and II. Edited by F. Rosengarten. Translated by R. Rosenthal. New York: Columbia University Press.

Greenberg, J. 2000. "Opinion Discourse and Canadian Newspapers: The Case of the Chinese Boat People." *Canadian Journal of Communication* 25, no. 4: 517–37.

Greenberg, J., and S. Hier. 2001. "Crisis, Mobilization and Collective Problematization: Illegal Migrants and the Canadian News Media. *Journalism Studies* 2, no. 4: 563–83.

Gregory, D. 2005. *The Colonial Present.* Malden, Mass.: Blackwell.

Grewal, I., and C. Kaplan, eds. 1994. *Scattered Hegemonies: Postmodernity and Transnational Feminist Practices.* Minneapolis: University of Minnesota Press.

Guarnizo, L., and M. Smith. 1998. "The Locations of Transnationalism." In *Transnationalism from Below*, edited by M. Smith and L. Guarnizo, 1–34. New Brunswick, N.J.: Transaction.

Guillet, E. C. 1963. *The Great Migration: The Atlantic Crossing by Sailing-Ship Since 1770.* Toronto: University of Toronto Press.

Gupta, A. 1995. "Blurred Boundaries: The Discourse of Corruption, the Culture of Politics, and the Imagined State." *American Ethnologist* 22, no. 2: 375–402.

Hage, G. 1998. *White Nation: Fantasies of White Supremacy in a Multicultural Society.* Annandale, N.S.W.: Pluto Press.

Hannah, M. 1997. "Imperfect Panopticism: Envisioning the Construction of Normal Lives." In *Space and Social Theory: Interpreting Modernity and Postmodernity,* edited by G. Benko and U. Strohmayer, 344–59. Oxford: Blackwell.

Hansen, T. B., and F. Stepputat. 2001. "Introduction: States of Imagination." In *States of Imagination: Ethnographic Explorations of the Postcolonial State,* edited by T. B. Hansen and F. Stepputat, 1–38. Durham, N.C.: Duke University Press.

———. 2005. *Sovereign Bodies: Citizens, Migrants, and States in the Postcolonial World.* Princeton, N.J.: Princeton University Press.

Haraway, D. 1988. "Situated Knowledges: The Science Question in Feminism and the Privilege of Partial Knowledge." *Feminist Studies* 14, no. 3: 575–99.

Harding, S. 1986. *The Science Question in Feminism.* Ithaca, N.Y.: Cornell University Press.

Hardt, M., and A. Negri. 2000. *Empire.* Cambridge, Mass.: Harvard University Press.

Harvey, D. 1990. *The Condition of Postmodernity.* Cambridge: Blackwell.

Herbert, S. 1997. *Policing Space: Territoriality and the Los Angeles Police Department.* Minneapolis: University of Minnesota Press.

———. 2000. "For Ethnography." *Progress in Human Geography* 24, no. 4: 550–68.

Herzfeld, M. 1992. *The Social Production of Indifference: Exploring the Symbolic Roots of Western Bureaucracy.* New York: Berg.

Heyman, J. M. 1995. "Putting Power in the Anthropology of Bureaucracy: The Immigration and Naturalization Service at the Mexico–United States Border." *Current Anthropology* 36, no. 2: 261–87.

———. 1999. "United States Surveillance over Mexican Lives at the Border: Snapshots of an Emerging Regime." *Human Organization* 58: 429–37.

———. 2007. "Grounding Immigrants Rights Movements in the Everyday Experience of Migration." *International Migration* 45, no. 3: 197–202.

Heyman, J. M., and A. Smart. 1999. "States and Illegal Practices: An Overview." In *States and Illegal Practices,* edited by J. Heyman, 1–24. New York: Berg.

Hier, S. P., and J. Greenberg. 2002. "Constructing a Discursive Crisis: Risk, Problematization and Illegal Chinese in Canada. *Ethnic and Racial Studies* 25, no. 3: 490–513.

Holquist, M. 2002. *Dialogism: Bakhtin and His World.* New York: Routledge.

Honig, B. 1998. "Immigrant America: How Foreignness "Solves" Democracy's Problems." *Social Text* 56: 1–27.

———. 2001. *Democracy and the Foreigner.* Princeton, N.J.: Princeton University Press.

Hood, M. 1997. "Sourcing the Problem: Why Fuzhou?" In *Human Smuggling,* edited by P. J. Smith, 76–92. Washington, D.C.: Center for Strategic & International Studies.

Howell, M. 2002. "People Smugglers: Organized Crime Rings Don't Always Use Rusty Freighters to Smuggle People—Hundreds of Migrants Slip into the Country Each Year by Air. *Vancouver Courier* 6 February.

Hugo, G. 2001. "From Compassion to Compliance? Trends in Refugee and Humanitarian Migration in Australia." *Geoforum* 55: 27–37.

Human Rights Watch. 2003. "An Unjust 'Vision" for Europe's Refugees." 17 June. http://hrw.org/backgrounder/refugees/uk (accessed 15 June 2004).

———. 2005. "Ukraine on the Margins: Rights Violations against Migrants and Asylum Seekers at the New Eastern Border of the European Union." *Human Rights Watch World Report* 17, no. 8.

Huysmans, J. 2006. *The Politics of Insecurity: Fear, Migration and Asylum in the EU.* New York: Routledge.

Hyndman, J. 2000. *Managing Displacement: Refugees and the Politics of Humanitarianism.* Minneapolis: University of Minnesota Press.

———. 2001. "Towards a Feminist Geopolitics." *Canadian Geographer* 45, no. 2: 210–22.

———. 2004. "Mind the Gap: Bridging Feminist and Political Geography through Geopolitics." *Political Geography* 23: 307–22.

Hyndman, J., and A. Mountz. 2006. "Refuge or Refusal: The Geography of Exclusion." In *Violent Geographies,* edited by D. Gregory and A. Pred, 77–92. New York: Routledge.

International Organization for Migration. 1997. "Trafficking in Migrants: IOM Policy and Activities." http://www.iom.ch/IOM/Trafficking/IOM_Policy.html (accessed 12 June 2003).

Jessop, B. 1990. "The State as Political Strategy." In *State Theory: Putting Capitalist States in Their Place,* 248–72. Cambridge: Polity Press.

Jiménez, M. 2001. "Going to America: Why the Truth about Illegal Chinese Migrants Isn't What You Think." *Saturday Night* 27 January.

Johnston, H. 1989. *The Voyage of the Komagata Maru: The Sikh Challenge to Canada's Colour Bar.* Vancouver: University of British Columbia Press.

Kalyvas, A. 2002. "The Stateless Theory: Poulantzas's Challenge to Postmodernism." In *Paradigm Lost: State Theory Reconsidered,* edited by S. Aronowitz and P. Bratsis, 105–42. Minneapolis: University of Minnesota Press.

Katz, C. 1992. "All the World Is Staged: Intellectuals and the Projects of Ethnography." *Environment and Planning D: Society and Space* 10: 495–510.

Kearney, M. 1991. "Borders and Boundaries of State and Self at the End of Empire." *Journal of Historical Sociology* 4, no. 1: 52–74.

Kerner C., A. Bailey, A. Mountz, I. Miyares, and R. Wright. 2001. "'Thank God She's Not Sick': Health and Disciplinary Practice among Salvadoran Women in Northern New Jersey." In *Geographies of Women's Health*, edited by I. Dyck, N. Lewis, and S. McLafferty, 127–42. New York: Routledge.

Klein, N. 2006. *The Shock Doctrine*. New York: Picador.

Koser, K. 2000. "Asylum Policies, Trafficking and Vulnerability." *International Migration* 38: 91–111.

———. 2001. "The Smuggling of Asylum Seekers into Western Europe: Contradictions, Conundrums, and Dilemmas." In *Global Human Smuggling: Comparative Perspectives*, edited by D. Kyle and R. Koslowski, 58–73. Baltimore: Johns Hopkins University Press.

Koslowski, R. 2005. "Smart Borders, Virtual Borders or No Borders: Homeland Security Choices for the United States and Canada." *Law and Business Review of the Americas* 11, no. 3/4: 527–50.

Kristeva, J. 1982. *Powers of Horror: An Essay on Abjection*. New York: Columbia University Press.

Kwong, P. 1997. *Forbidden Workers: Illegal Chinese Immigrants and American Labor.* New York: New York Press.

Kyle, D., and J. Dale. 2001. "Smuggling the State Back In: Agents of Human Smuggling Reconsidered." In *Global Human Smuggling: Comparative Perspectives*, edited by D. Kyle and R. Koslowski, 29–57. Baltimore: Johns Hopkins University Press.

Kyle, D., and R. Koslowski, eds. 2001. *Global Human Smuggling: Comparative Perspectives*. Baltimore: Johns Hopkins University Press.

Lahav, G. 1998. "Immigration and the State: The Devolution and Privatisation of Immigration Control in the EU." *Journal of Ethnic and Migration Studies* 24: 142–67.

Lahav, G., and V. Guiraudon. 2000. "Comparative Perspectives on Migration Control: Away from the Border and Outside the State." In *The Wall Around the West*, edited by P. Andreas and T. Snyder, 115–38. Lanham, Md.: Rowman & Littlefield.

Lawson, V. 2000. "Arguments within Geographies of Movement: The Theoretical Potential of Migrants' Stories." *Progress in Human Geography* 24, no. 2: 173–89.

Ley, D. 2003. "Seeking *Homo Economicus*: The Strange Story of Canada's Business Immigration Program." *Annals of the Association of American Geographers* 93, no. 3.

———. 2010. *Millionaire Migrants: Trans-Pacific Life Lines*. Oxford: Wiley-Blackwell.

Ley, D., and D. Hiebert. 2001. "Immigration Policy as Population Policy." *Canadian Geographer* 45, no. 1: 120–25.

Li, P. 1998. *Chinese in Canada*. Toronto: Oxford University Press.

Lonely Planet. 2003. *From Nothing to Zero: Letters from Refugees in Australia's Detention Centres*. Melbourne: Lonely Planet.

Longhurst, R. 2001. *Bodies: Exploring Fluid Boundaries*. New York: Routledge.

Lubiano, W. 1996. "Like Being Mugged by a Metaphor: Multiculturalism and State Narratives." In *Mapping Multiculturalism*, edited by A. Gordon and C. Newfield, 64–75. Minneapolis: University of Minnesota Press.

Macklin, A. 2003a. "Dancing across Borders: 'Exotic Dancers,' Trafficking, and Canadian Immigration Policy." *International Migration Review* 37, no. 2: 464–500.

———. 2003b. "The Value(s) of the Canada–US Safe Third Country Agreement." Caledon Institute of Social Policy. http://ssrn.com/abstract=557005 (accessed 26 July 2004).

———. 2005. "Disappearing Refuges: Reflections on the Canada-US Safe Third Country Agreement." Columbia Human Rights Law Review 36:365–426

Mahler, S. 1995. *American Dreaming: Immigrant Life on the Margins.* Princeton, N.J.: Princeton University Press.

———. 1998. "Theoretical and Empirical Contributions Toward a Research Agenda for Transnationalism." In *Transnationalism from Below*, edited by M. Smith and P. Guarnizo, 64–94. New Brunswick, N.J.: Transaction.

Mahtani, M., and A. Mountz. 2002. "Immigration to British Columbia: Media Representation and Public Opinion." Research on Immigration and Integration in the Metropolis Working Paper No. 02-15. http://www.riim.metropolis.net/ (accessed 1 May 2003).

Marchand, M., and A. S. Runyan, eds. 2000. *Gender and Global Restructuring: Sightings, Sites and Resistances.* New York: Routledge.

Mares, P. 2002. *Borderline: Australia's Response to Refugees and Asylum Seekers.* Sydney: UNSW Press.

Maril, R. 2004. *Patrolling Chaos: The U.S. Border Patrol in Deep South Texas.* Lubbock: Texas Tech University.

Marr, D. 2009. "The Indian Ocean Solution." *The Monthly.* http://www.themonthly.com.au/print/1940 (accessed 29 December 2009).

Marston, S. A. 2000. "The Social Construction of Scale. *Progress in Human Geography* 24, no. 2: 219–42.

Massey, D., J. Durand, and N. Malone. 2002. *Beyond Smoke and Mirrors: Mexican Immigration in an Era of Economic Integration.* New York: Russell Sage Foundation.

Mayer, J. 2005. "Outsourcing Torture." *New Yorker* 14 February.

McBride, D. 1999. "Migrants and Asylum Seekers: Policy Responses in the United States to Immigrants and Refugees from Central American and the Caribbean." *International Migration* 37, no. 1: 289–317.

McGuinness, S. 2001. "Canadian Print Media Coverage of the 1999 Fujian Migrants." Centre of Chinese Research, Institute of Asian Research, University of British Columbia.

McGuirk, R. 2007. "Australia, U.S. to Transfer Refugees." Immigration Watch Canada. 18 April 2008. http://www.immigrationwatchcanada.org/index.php?PAGE_id=1592&PAGE_user_op=view_page&module=pagemaster (accessed 3 January 2010).

McHugh, K. 2000. "Inside, Outside, Upside Down, Backward, Forward, Round and Round: A Case for Ethnographic Studies in Migration." *Progress in Human Geography* 24, no. 1: 71–89.

Menjívar, C. 2000. *Fragmented Ties: Salvadoran Immigrant Networks in America.* Berkeley: University of California Press.

Migration News. 2006. "Canada: Unauthorized, Asylum." http://migration.ucdavis.edu/mn/comments.php?id=3182_0_2_0 (accessed 28 February 2009).

Mitchell, K. 1993. "Multiculturalism, or the United Colors of Capitalism?" *Antipode* 25, no. 4: 263–94.

———. 1997a. "Different Diasporas and the Hype of Hybridity." *Environment and Planning D: Society and Space* 15: 533–53.

———. 1997b. "Transnational Discourse: Bringing Geography Back In." *Antipode* 29, no. 2: 101–14.

———. 2006. "Geographies of Identity: The New Exceptionalism." *Progress in Human Geography* 30, no. 1: 95–106.

Mitchell, T. 1991. "The Limits of the State: Beyond Statist Approaches and Their Critics." *American Political Science Review* 85, no. 1: 77–96.

Miyares, I. 2003. "The Interrupted Circle: Truncated Transnationalism and the Salvadoran Experience." *Journal of Latin American Geography* 2, no. 1: 74–86.

Moorehead, C. 2005. *Human Cargo: A Journey among Refugees.* New York: Picador/ Henry Holt.

Morris, M. D. 1985. *Immigration: The Beleaguered Bureaucracy.* Washington, D.C.: Brookings Institution.

Mountz, A. 1995. "Daily Life in a Transnational Community: The Fusion of San Agustín, Oaxaca and Poughkeepsie, New York." Thesis, Senior Fellowship, Dartmouth College.

———. 2003. "Human Smuggling, the Transnational Imaginary, and Everyday Geographies of the Nation-State." *Antipode* 35, no. 3: 622–44.

———. 2004. "Embodying the Nation-State: Canada's Response to Human Smuggling." *Political Geography* 23: 323–45.

———. 2006. "Human Smuggling and the Canadian State." *Canadian Foreign Policy* 13, no. 1: 59–80.

———. 2007. "Smoke and Mirrors: An Ethnography of State." In *Politics and Practice in Economic Geography,* edited by E. Sheppard, T. Barnes, J. Peck, and A. Tickell, 38–48. Thousand Oaks, Calif.: Sage.

Mountz, A., and R. Wright. 1996. "Daily Life in the Transnational Migrant Community of San Agustín, Oaxaca and Poughkeepsie, New York." *Diaspora* 5, no. 3: 403–28.

Mountz, A., R. Wright, I. Miyares, and A. Bailey. 2002. "Lives in Limbo: Temporary Protected Status and Immigrant Identities." *Global Networks* 2, no. 4: 335–56.

Nader, L. 1972. "Up the Anthropologist: Perspectives Gained from Studying Up." In *Reinventing Anthropology,* edited by D. Hymes, 284–311. New York: Random House.

Nadig, A. 2002. "Human Smuggling, National Security, and Refugee Protection." *Journal of Refugee Studies* 15, no. 1: 1–25.

Nagar, R., V. Lawson, L. McDowell, and S. Hanson. 2002. "Locating Globalization: Feminist (Re)readings of the Subjects and Spaces of Globalization." *Economic Geography* 78, no. 3: 257–84.

Nagel, J. 2003. *Race, Ethnicity, and Sexuality: Intimate Intersections, Forbidden Frontiers.* New York: Oxford University Press.

Nast, H. 1998. "Unsexy Geographies." *Gender, Place and Culture* 5, no. 2: 191–206.

National Post. 2001. "Alleged Terrorist's Canadian Trial." 13 March.

Nelson, D. 1999. *A Finger in the Wound: Body Politics in Quincentennial Guatemala.* Berkeley: University of California Press.

———. 2004. "Anthropologist Discovers Legendary Two-Faced Indian! Margins, the State, and Duplicity in Postwar Guatemala." In *Anthropology in the Margins of the State,* edited by V. Das and D. Poole, 117–40. New York: Oxford University Press.

Nevins, J. 2002. *Operation Gatekeeper.* New York: Routledge.

Nevins, J., and M. Aizeki. 2008. *Dying to Live: A Story of U.S. Immigration in an Age of Global Apartheid.* San Francisco: City Lights Open Media.

Newland, K. 2005. "Drop in Asylum Numbers Shows Changes in Demand and Supply." Migration Information Source, Migration Policy Institute. http://www.migrationinformation.org/Feature/print.cfm?ID=303 (accessed 2 June 2005).

New York Times. 2000a. "U.N. Warns That Trafficking in Human Beings Is Growing." 25 June.

———. 2000b. "U.S. Seeks China's Help in Slowing the Flood of Illegal Immigrants." 4 July.

———. 2001. "Stowaway Fell from Jet Near Airport, Police Say." 9 August.

———. 2002. "Canada Courts Migrant Families to Revive a Declining Hinterland." 2 October.

———. 2003. "Hong Kong Human Smuggler Gets Four Years." 8 February.

———. 2007. "Gitmos across America." 27 June.

Nichol, H. 2005. "Resiliency or Change? The Contemporary Canada–US Border." *Geopolitics* 10, 767–90.

Nyers, P. 2006. *Rethinking Refugees: Beyond States of Emergency.* New York: Routledge.

Ohmae, K. 1995. *The End of the Nation State: The Rise of Regional Economies.* New York: Free Press.

Okólski, M. 2000. "Illegality of International Population Movements in Poland." *International Migration* 38: 57–87.

Ong, A. 1999. *Flexible Citizenship: The Cultural Logics of Transnationality.* Durham, N.C.: Duke University Press.

———. 2006. *Neoliberalism as Exception: Mutations in Citizenship and Sovereignty.* Durham, N.C.: Duke University Press.

Painter, Joe. 1995. *Politics, Geography, and "Political Geography."* New York: Arnold.

———. 2006. "Prosaic Geographies of Stateness." *Political Geography* 25, no. 7: 752–74.

Palmer, D. 1998. "A Detailed Regional Analysis of Perceptions of Immigration in Canada." Submitted to Strategic Policy, Planning and Research, Citizenship and Immigration Canada.

Paucurar, A. 2003. "Smuggling, Detention and Expulsion of Irregular Migrants: A Study on International Legal Norms, Standards and Practices." *European Journal of Migration and Law* 2: 259–83.

Perera, S. 2002a. "A Line in the Sea." *Race & Class* 44, no. 2: 23–39.

———. 2002b. "The Tampa, Boat Stories and the Border: A Line in the Sea." *Cultural Studies Review* 8, no. 1: 11–27.

———. 2002c "What Is a Camp . . . ?" *Borderlands* e-journal 1, no. 1. http://www .borderlands.net.au/vol1no1_2002/perera_camp.html (accessed 2 January 2010).

Portes, A., L. Guarnizo, and P. Landolt. 1999. "The Study of Transnationalism: Pitfalls and Promise of an Emergent Research Field." *Ethnic and Racial Studies* 22, no. 2: 217–23.

Pratt, A. 2005. *Securing Borders: Detention and Deportation in Canada.* Vancouver: University of British Columbia Press.

Pratt, A., and S. Thompson. 2008. "Chivalry, 'Race' and Discretion at the Canadian Border." *British Journal of Criminology* 48: 620–40.

Pratt, G. 1999. "From Registered Nurse to Registered Nanny: Discursive Geographies of Filipina Domestic Workers in Vancouver, B.C. *Economic Geography* 75, no. 3: 215–36.

———. 2000. "Research Performances." *Environment and Planning D: Society and Space* 18, no. 5: 639–51.

———. 2004. *Working Feminism.* Philadelphia: Temple University Press.

Pratt, G., in collaboration with the Philippine Women's Centre. 1998. "Inscribing Domestic Work on Filipina Bodies." In *Places through the Body*, edited by H. Nast and S. Pile, 283–304. London: Routledge.

Pred, A. 1995. *Even in Sweden.* Berkeley: University of California Press.

Probyn, E. 1996. *Outside Belongings.* New York: Routledge.

Province. 1999a. "Quarantined." 21 July.

———. 1999b. "Beware, Illegal Immigrants. We Canadians Can Be Pretty Ruthless." 13 August.

———. 1999c. "Enough Already. It's Time to Toughen the Law." 1 September.

Puar, J. 2007. *Terrorist Assemblages: Homonationalism in Queer Times.* Durham, N.C.: Duke University Press.

Purcell, M., and J. Nevins. 2005. "Pushing the Boundary: State Restructuring, State Theory, and the Case of U.S.–Mexico Border Enforcement in the 1990s." *Political Geography* 24: 211–35.

Putnam, R. 2000. *Bowling Alone: The Collapse and Revival of American Community.* New York: Schuster.

Rajaram, P. K. 2002. "Humanitarianism and Representations of the Refugee." *Journal of Refugee Studies* 15, no. 3: 247–64.

Razack, S. 1999. "Making Canada White: Law and the Policing of Bodies of Colour in the 1990s." *Canadian Journal of Law & Society* 14, no. 1: 159–84.

Rivers and Associates. 2000. "Opinion Polls as Baseline Measures: A Discussion Paper." Prepared for the Ministry of Multiculturalism and Immigration, Province of British Columbia.

Rose, G. 1993. *Feminism & Geography: The Limits of Geographical Knowledge.* Minneapolis: University of Minnesota Press.

———— 1995. "Geography and Gender, Cartographies and Corporealities." *Progress in Human Geography* 19, no. 4: 544–48.

Rouse, R. 1991. "Mexican Migration and the Social Space of Postmodernism." *Diaspora* 1: 8–23.

————. 1992. "Making Sense of Settlement: Class Transformation, Cultural Struggle, and Transnationalism among Mexican Migrants in the United States." In *Towards a Transnational Perspective on Migration: Race, Class, Ethnicity, and Nationalism Reconsidered,* edited by N. Glick-Schiller, L. Basch, and C. Blanc-Szanton, 26–47. New York: New York Academy of Sciences.

————. 1995. "Thinking through Transnationalism: Notes on the Cultural Politics of Class Relations in the Contemporary United States." *Public Culture* 7: 353–402.

Rutvica, A. 2006. "Lampedusa in Focus: Migrants Caught between the Libyan Desert and the Deep Sea." COMPAS Policy Documentation Center Working Paper. http://pdc.ceu.hu/archive/00002836/04/lampedusa.pdf (accessed 4 June 2006).

Salt, J., and J. Stein. 1997. "Migration as a Business: The Case of Trafficking." *International Migration* 35, no. 4: 467–94.

Salter, M. 2004. "Passports, Mobility, and Security: How Smart Can the Border Be?" *International Studies Perspectives* 5: 71–91.

————. 2008. "Governmentalities of an Airport: Heterotopia and Confession." *International Studies* 1, no. 1: 49–66.

Sanchez, L. 2004. "The Global E-rotic Subject, the Ban, and the Prostitute-Free Zone: Sex Work and the Theory of Differential Exclusion." *Environment and Planning D: Society and Space* 22, no. 6: 861–83.

Sassen, S. 1996. *Losing Control: Sovereignty in an Age of Globalization.* New York: Columbia University Press.

————. 2006. *Territory, Authority, Rights.* Princeton, N.J.: Princeton University Press.

Schuster, L. 2005a. "A Sledgehammer to Crack a Nut: Deportation, Detention, and Dispersal in Europe." *Social Policy & Administration* 39, no. 6: 606–21.

————. 2005b. "The Realities of a New Asylum Paradigm." COMPAS Working Paper No. 20. Oxford: University of Oxford, Policy Documentation Center.

Scott, J. 1998. *Seeing Like a State: How Certain Schemes to Improve the Human Condition Have Failed.* New Haven, Conn.: Yale University Press.

Sharma, A., and A. Gupta, eds. 2006. *An Anthropology of the State: A Reader*. Oxford: Blackwell.

Sharma, N. 2001. "On Being Not Canadian: The Social Organization of 'Migrant Workers' in Canada." *Canadian Review of Sociology and Anthropology* 38, no. 4: 415–39.

———. 2005. "Anti-Trafficking Rhetoric and the Making of a Global Apartheid." *NWSA Journal* 17, no. 3: 88–108.

Sibley, D. 1995. *Geographies of Exclusion*. New York: Routledge.

Silvey, R., and V. Lawson. 1999. "Placing the Migrant." *Annals of the Association of American Geographers* 89, no. 1: 121–32.

Simon, J. 1998. "Refugees in a Carceral Age: The Rebirth of Immigration Prisons in the United States." *Public Culture* 10, no. 3: 577–607.

Singh, N. 1994. *Canadian Sikhs: History, Religion, and Culture of Sikhs in North America*. Ottawa: Canadian Sikhs' Studies Institute.

Skeldon, R. 2000a. *Myths and Realities of Chinese Irregular Migration*. IOM Migration Research Series, No. 1. Geneva: International Organization for Migration.

———. 2000b. "Trafficking: A Perspective from Asia." *International Migration* 38: 8–28.

Smith, D. 1987. *The Everyday World as Problematic: A Feminist Sociology*. Toronto: University of Toronto Press.

———. 1990. *The Conceptual Practices of Power: A Feminist Sociology of Knowledge*. Boston: Northeastern University Press.

Smith, N. 2001. "A Commentary on Events in New York." http://www.peoplesgeography .org/Smith.html (accessed 1 June 2003).

Smith, P. 1997. "Chinese Migrant Trafficking: A Global Challenge." In *Human Smuggling*, edited by P. Smith, 1–22. Washington, D.C.: Center for Strategic & International Studies.

Soysal, Y. N. 2000. "Citizenship and Identity: Living in Diasporas in Post-war Europe?" *Ethnic and Racial Studies* 23, no. 1: 1–15.

Sparke, M. 2005. *In the Space of Theory: Postfoundational Geographies of the Nation-State*. Minneapolis: University of Minnesota Press.

———. 2006. "A Neoliberal Nexus: Citizenship, Security and the Future of the Border." *Political Geography* 25, no. 2: 151–80.

Spencer, D. 2000. "The Logic and Contradictions of Intensified Border Enforcement in Texas." In *The Wall around the West*, edited by P. Andreas and T. Snyder, 115–38. Lanham, Md.: Rowman & Littlefield.

Steel, Z., D. Silove, R. Brooks, S. Momartin, B. Alzuhairi, and I. Suslijik. 2006. "Impact of Immigration Detention and Temporary Protection on the Mental Health of Refugees." *British Journal of Psychiatry* 188: 58–64.

Steinmetz, G. 1999. "Introduction: Culture and the State." In *State/Culture: State-Formation after the Cultural Turn*, edited by G. Steinmetz, 1–49. Ithaca, N.Y.: Cornell University Press.

Taussig, M. 1997. *The Magic of the State*. New York: Routledge.

Taylor, P. 2000. "Embedded Statism and the Social Sciences 2: Geographies (and Metageographies) in Globalization." *Environment and Planning A* 32: 105–14.

Telegraph. 2006. "Adopt Our Values or Stay Away, Says Blair." 9 December.

Thrift, N. 2000. "It's the Little Things." In *Geopolitical Traditions: A Century of Geopolitical Thought,* edited by K. Dodds and D. Atkinson, 380–87. New York: Routledge.

Torpey, J. 2000. *The Invention of the Passport: Surveillance, Citizenship and the State.* Cambridge: Cambridge University Press.

Tyner, J. 2000. "Migrant Labour and the Politics of Scale: Gendering the Philippine State." *Asia Pacific Viewpoint* 41, no. 2: 131–54.

United Nations. 2000. "Protocol to Prevent, Suppress, and Punish Trafficking in Persons, Especially Women and Children, Supplementing the United Nations Convention against Transnational Organized Crime." G.A. Res. 55/25, Annex II, 55 UN GAOR Supp. (no. 49) at 60, UN Doc A/45/49 (vol. 1) (2001). http://www.uncjin.org/Documents/Conventions/dcatoc/final_documents_2/convention_%20traff_eng.pdf. (accessed 17 March 2003).

United Nations High Commissioner for Refugees. 2000. "UNHCR Summary Position on the Protocol against the Smuggling of Migrants by Land, Sea and Air and the Protocol to Prevent, Suppress and Punish Trafficking in Persons, Especially Women and Children, Supplementing the UN Convention against Transnational Organized Crime." 11 December.

———. 2005. "UNHCR Deeply Concerned over Lampedusa Deportations." Press release, 18 March. www.unhcr.ch/press-releases (accessed 20 March 2005).

———. 2007. *2006 Global Trends.* Geneva: UNHCR.

United States Committee for Refugees. 1998a. "Asylum Cases Filed with the INS." *Refugee Reports* 19: 6.

———. 1998b. "Country Report: United States." http://www.refugees.org/world/countryrpt/amer_carib/1998/us.htm (accessed 25 April 2003).

———. 1999. "Where America's Day Begins: Chinese Asylum Seekers on Guam." http://www.refugees.org/world/articles/asylum_rr99_8.htm (accessed 1 June 2003).

———. 2000a. "Country Report: Canada." http://www.refugees.org/world/countryrpt/amer_carib/2000/canada.htm (accessed 23 January 2003).

———. 2000b. "NGOs Call on UNHCR Executive Committee to Oppose Interception." http://www.refugees.org/world/articles/ngo_rr00_5.htm (accessed 1 June 2003).

———. 2001. "Country Report: Canada." http://www.refugees.org/world/countryrpt/amer_carib/2001/canada.htm (accessed 23 January 2003).

———. 2003. "United Kingdom: World Refugee Survey 2003 Country Report." http://www.uscr.org/world/countryrpt/europe/2003/united_kingdom.cfm (accessed 14 June 2004).

United States Committee for Refugees and Immigrants. 2009. "World Refugee Survey: 2009." http://www.refugees.org/article aspx?id=2324&subm=179&area=About%20Refugees& (accessed 4 January 2010).

United States Department of Transportation. 2005. "Smart Border Action Plan, Status Report." http://www.fhwa.dot.gov/uscanada/issues/action_plan.thm (accessed 28 September 2005).

Valentine, David. 2007. *Imagining Transgender: An Ethnography of a Category.* Durham, N.C.: Duke University Press.

Van Impe, K. 2000. "People for Sale: The Need for a Multidisciplinary Approach towards Human Trafficking." *International Migration* 38: 113–32.

Vancouver Sun. 1999. "Detained Aliens Investigated." 22 July.

———. 2000. "Canada to Blame for Migrants, China Says." 7 February.

———. 2009. "10 Years After: B.C.'s Chinese Boat Migrants." 20 October.

Vila, P., ed. 2003. *Ethnography at the Border.* Minneapolis: University of Minnesota Press.

Vis-à-vis Millenium. 2000. Citizenship and Immigration Canada, B.C.–Yukon Region: 1–3. January/February.

Waldinger, R., and D. Fitzgerald. 2004. "Transnationalism in Question." *American Journal of Sociology* 5: 1177–95.

Walters, W. 2002. "Deportation, Expulsion, and the International Police of Aliens." *Citizenship Studies* 6, no. 3: 265–81.

———. 2004. "Secure Borders, Safe Haven, Domopolitics." *Citizenship Studies* 8, no. 3: 237–60.

Walton-Roberts, M. 2001. "Embodied Global Flows: Immigration and Transnational Networks between British Columbia, Canada, and Punjab, India." Department of Geography, University of British Columbia.

Wasem, R. E., and K. Ester. 2004. "Temporary Protected Status: Current Immigration Policy and Issues." CRS Report for Congress. Congressional Research Service, Library of Congress. http://fpc.state.gov/documents/organization/41113.pdf (accessed 24 February 2009).

Washington Times. 2007. "U.S. to Ship Refugees Half a World Away." 19 April. http://www.washtimes.com/world/20070418-100355-4106r.htm (accessed 19 April 2007).

Weber, M. 1947. *The Theory of Social and Economic Organization.* New York: Oxford University Press.

Welch, M. 2002. *Detained: Immigration Laws and the Expanding I.N.S. Jail Complex.* Philadelphia: Temple University Press.

Wolch, J. 1989. "The Shadow State: Transformations in the Voluntary Sector." In *The Power of Geography,* edited by J. Wolch and M. Dear, 197–221. Boston: Unwin Hyman.

Wright R., A. Bailey, I. Miyares, and A. Mountz. 2000. "Legal Status, Gender and Employment among Salvadorans in the US." *International Journal of Population Geography* 6: 273–86.

Yates, K. 1997. "Canada's Growing Role as a Human Smuggling Destination and Corridor to the United States." In *Human Smuggling,* edited by P. J. Smith, 156–68. Washington, D.C.: Center for Strategic & International Studies.

Index

Access to Information Act, 20, 82–84
Airline Liaison Officers, 42, 127, 137.
  See also Immigration Control Officers;
  interdiction
Amelie (motor vessel), 8, 36
antiterrorism legislation, 51, 142, 144,
  173
Australia: activists in, 169; detention,
  131–32, 135; influence on Canada, 1,
  8, 10, 47, 48, 126; offshore enforcement 49, 124, 127–29, 136, 138–40,
  174; "policy on the run," 20. *See also*
  islands; Pacific Solution

Bill C-11, 10, 17
Bill C-36, 51

Canadian Border Services Agency
  (CBSA): creation, 41, 177, 181; mandate, 32, 43, 49–51, 125, 185
Canadian Charter of Rights and Free-
  doms, 16, 74
chartered flights, 87, 97, 98, 169, 173
citizen: good, 78, 161; new, 61, 65,
  172; racialization of, 150–51; rights,
  144, 158; skilled, xvi; state and, 58,
  90; undocumented, xxix, 163, 164;
  wealthy, xvi, 101. *See also* surveillance
citizenship: belonging and, xxviii, 159
  184n19; ceremonies, 61, 65; mobility

and, 138, 148, 152; scholarship on,
  xxii; state and, 101; temporary, 157.
  See also citizen; Temporary Protected
  Status (TPS)
Citizenship and Immigration Canada
  (CIC): access to information, 82–84,
  94; access to refugee determination
  process and, 103–4, 109, 180; budget, 61; communications, 63, 71, 74;
  detention, 85, 118, 183; Immigration Control Officers, 48–49; interceptions, 1, 11, 14, 15; long tunnel thesis
  and, xiii, xiv; mandates, 15, 84, 185;
  media, 68–69, 72, 98; research with,
  xvii, xix, xx, xxv, 171, 177n3; staffing,
  61, 65. *See also* intelligence; policy
civil society: "keep watch" programs,
  33, 34, 44, 52, 165; politicization
  of, 178n6; state and, xxiv; structural
  effect and, xxi. *See also* surveillance
criminalization: and access, 106; of asylum seekers, 124, 172–74; discursive,
  113, 130; of disorderly movement, 78,
  122; through homogenization, 142–44;
  media representations of, 103, 105,
  182n2; of migration xvi, xxvii, 12, 141,
  156, 163; post-traumatic stress disorder
  (PTSD) and, 159; self-surveillance and,
  151; surveillance and, 164. *See also*
  media coverage; racialization

205

ALISON MOUNTZ is associate professor of geography at Syracuse University. She is 2009–2010 William Lyon Mackenzie King Research Fellow with the Canada Program at Harvard University. Her latest research examines island detention centers off the shores of North America, Australia, and the European Union and is funded by a CAREER grant from the National Science Foundation.